Please return/renew this item by the last date shown on this label, or on your self-service receipt.

To renew this item, visit **www.librarieswest.org.uk** or contact your library.

Your Borrower number and PIN are required.

# How Evolution Explains Everything About Life

*From Darwin's brilliant idea to today's epic theory*

NEW SCIENTIST

First published in Great Britain in 2017 by John Murray Learning
First published in the US in 2017 by Nicholas Brealey Publishing
Copyright © *New Scientist* 2017

*A catalogue record for this book is available from the British Library and the Library of Congress.*
UK Paperback ISBN 978 1473 62971 4 / eBook ISBN 978 1473 62972 1
US ISBN 978 1473 65845 5 / eISBN 978 1473 65844 8
1

Typeset by Cenveo® Publisher Services.
Printed and bound in Great Britain by CPI Group (UK) Ltd., Croydon CR0 4YY
Hachette UK's policy is to use papers that are natural, renewable and recyclable products and
made from wood grown in sustainable forests. The logging and manufacturing processes are
expected to conform to the environmental regulations of the country of origin.

Carmelite House
50 Victoria Embankment
London
EC4Y 0DZ
www.hodder.co.uk

Nicholas Brealey Publishing
Hachette Book Group
53 State Street
Boston MA 02109
www.nicholasbrealey.com

# Contents

# Series introduction

*New Scientist's Instant Expert* books shine light on the subjects that we all wish we knew more about: topics that challenge, engage enquiring minds and open up a deeper understanding of the world around us. *Instant Expert* books are definitive and accessible entry points for curious readers who want to know how things work and why. Look out for the other titles in the series:

*The End of Money*
*How Your Brain Works*
*Machines That Think*
*The Quantum World*
*Where the Universe Came From*
*Your Conscious Mind*

# Contributors

This book is based on articles previously published in *New Scientist* together with specially commissioned content. It is authored by a range of experts

Editor: **Alison George**, *Instant Expert* editor for *New Scientist*. Alison has a PhD in biochemistry and worked as a microbiologist with the British Antarctic Survey.

**Adrian Bird** is professor of genetics at the University of Edinburgh. He wrote about epigenetics in Chapter 8.

**Sue Blackmore** is a psychologist, lecturer and writer researching consciousness, memes and anomalous experiences, and is a visiting professor at the University of Plymouth. She wrote about memes in Chapter 3.

**Peter Bowler** is a historian of science specializing in evolution, and emeritus professor at Queens University Belfast. He wrote about the genetic revolution of evolution in Chapter 3.

**Lee Alan Dugatkin** is professor of biological sciences at the University of Louisville. He wrote 'Taking it personally' in Chapter 9.

**Steve Jones** is emeritus professor of human genetics at University College London. He wrote '*On the Origin of Species*' Revisited, Chapter 11.

**Kevin Laland** is professor of behavioural and evolutionary biology at the University of St Andrews. He wrote 'Evolution evolves: Beyond the selfish gene' in Chapter 10.

**George Turner** is Lloyd Roberts Professor of Zoology at Bangor University, specializing in cichlid fish biology and evolution. He wrote about how new species form in Chapter 7.

**David Sloan Wilson** is professor of biological sciences and anthropology at Binghamton University. He wrote about group selection in Chapter 9.

**John van Wyhe** is a historian of science at the National University of Singapore and founder of Darwin Online (darwin-online.org.uk/). He wrote about how Darwin and Wallace came up with their theory of evolution in Chapter 1.

**Thanks also to the following writers and editors:**

Claire Ainsworth, Colin Barras, Michael Brooks, Mark Buchanan, Michael Chorost, Bob Holmes, Rowan Hooper, Dan Jones, Simon Ings, Shelly Innes Graham Lawton, Michael le Page, Alison Pearn, Paul Rainey, Penny Sarchet, John Waller.

x

# Introduction

'One general law, leading to the advancement of all organic beings, namely, multiply, vary, let the strongest live and the weakest die.'

So wrote Charles Darwin in 1859 in *On the Origin of Species.* In this book, he outlined why life is so diverse, and living things are so well suited to their environments: evolution by natural selection.

We now know how, over 3.8 billion years, evolution's blind, brutish and aimless methods filled a once-barren planet with the rich diversity of plants, animals, fungi and microbes that surround us. We understand how simple processes can produce astonishingly complex structures, from wings and eyes to biological computers and solar panels.

But Darwin, together with Alfred Russel Wallace — who devised his theory of evolution at the same time — did more than explain the diversity of life. They also upended humanity's view of itself as a special creation of god, showing that we are just one tiny branch on a vast tree of life, all descended from a common ancestor.

Time has not diminished these insights, but it has added to them. Another revolution was triggered in the 1930s and 1940s when the new science of genetics was incorporated into the theory of evolution. We now understand evolution in terms of the propagation of genes.

This guide explores the inner workings of evolution, and addresses the tricky questions it raises. Was life inevitable or a one-off fluke, and how did it get kick-started? Does it have a purpose or direction? The book also highlights life's greatest inventions (and mistakes) and investigates the thorny issue of how altruistic traits evolve, still a live issue more than 150 years after Darwin first discussed it.

Although Darwin and Wallace's fundamental idea of evolution by natural selection has stood the test of time, some biologists argue that our theory of life needs another revolution, as we discover more about the complexities of how evolution works.

*How Evolution Explains Everything About Life* brings you up to date with the past, present and future of the science of evolution, and its intriguing implications.

**Alison George**

# I
# Darwin's discovery

*Not long ago, it was thought that God created all species. That changed in 1858 when the work of Charles Darwin and Alfred Russel Wallace showed, irrefutably, that humans are just another animal occupying a small branch on a vast tree of life. How exactly did Darwin devise his theory of evolution? What ideas did he build on? Was Wallace robbed of the credit? And how shocking was the idea to the Christian society of the time?*

## *The evolution revolution*

Traditionally, people in Christian Europe had believed that the world was about 6000 years old. This view was guided by interpolations from the Bible, which itself gave no date for creation. Gradually such beliefs were modified by Christian thinkers based on new information about Earth, gleaned from the growth of mining and the development of geology. By the early nineteenth century it was widely understood that Earth could not be a few thousand years old, but must be inconceivably ancient.

Earth was also found to have changed over time. Close study of rocks and fossils revealed a complex history of different ages. One layer of the geological record might show lush tropical vegetation populated with reptiles unlike any alive today. In the rock layers just above, yet another terrestrial world might have existed with different animals and plants. To explain this, in 1812 the great French anatomist Georges Cuvier put forward the idea that each age had been abruptly ended by some great catastrophe.

Another puzzle was the discovery in Europe and America of gigantic fossilized animals. Where could creatures such as mammoths be living today? Perhaps their kind had died out? This couldn't be true, according to traditional belief, since God would not allow any of his created species to perish.

Cuvier's detailed research in anatomy established, once and for all, that creatures such as the mammoth were not the same as anything alive today, and were extinct. For us, extinction is such a mundane fact that we cannot appreciate how radical the concept was initially. However, it soon became almost universally accepted in the scientific community, with one important exception: the French naturalist Jean-Baptiste Lamarck.

## Lamarckism

For Lamarck, these unfamiliar fossil forms had not become extinct. Instead they had changed, evolving into something else – although his view of this process was different to that later proposed by Darwin (see 'The view before Darwin'). The mammoth, for example, could have evolved into the elephant.

---

### The view before Darwin

Lamarckism is mentioned in biology textbooks as shorthand for a pre-Darwinian theory of evolution in which species were thought to evolve via the inheritance of characteristics acquired during an organism's lifetime. According to this, the giraffe got its long neck by intentionally stretching to reach the top of trees. This slightly stretched neck was then passed on to offspring, and so on. But this was not the core of his theory. Instead, Lamarck's central idea was that there is a tendency for life to progress up the scale of perfection according to a 'complexifying force'. New species originated continuously via spontaneous generation (the appearance of life from inorganic matter) and could adapt to local circumstances through the inheritance of acquired characteristics. Many naturalists, including Darwin, continued to accept Lamarck's ideas about evolution.

---

As the influential Cuvier did with so many rivals, he used his reputation to demolish Lamarck. The result was that for the first few decades of the nineteenth century, not only Lamarck's theory but any theory of evolution was considered unscientific and absurd. Although Lamarck won a few converts, many more accepted Cuvier's view that a succession of eras of life had come and gone.

But where did the new species that emerged after these extinctions come from? The geologist Charles Lyell argued in his *Principles of Geology*, published in the early 1830s, that slow processes had changed Earth over time. Lyell's picture was one of perpetual change. As an environment gradually transformed, the species that lived in it would become unsuited to it and die out, because there was a limit to how much they could change to adapt. Just how new species arose was left vague.

Lyell's work was of great interest to Charles Darwin, a young Cambridge graduate who was appointed to join the surveying voyage of HMS *Beagle* in 1831 as a naturalist (contrary to

FIGURE 1.1 Most pictures of Charles Darwin were taken when he was an old man, but he was only 22 years old when he set sail on the *Beagle*.

popular accounts, he was not invited along to be the captain's social companion, nor was the ship's surgeon the official naturalist). During this five-year voyage, Darwin matured into one of the most experienced scientists of his generation. He worked primarily as a geologist but also collected a wide range of living things, from finches to fungi.

The expedition first visited South America, then surveyed the waters around the Galapagos Islands. Only in the middle of the twentieth century did Darwin's visit there come to be portrayed as a pivotal moment in his life. He never described it as such. And as charming as it sounds, there is no truth to the story that Darwin noticed the beaks of the finches were adapted to different diets and that this provoked his evolutionary theorizing. There was no Galapagos eureka moment.

## Deep questions about nature

After the return of HMS *Beagle* in 1836, Darwin set to work describing his mountain of specimens. He also began asking himself deep questions about nature, life and religion. Gradually, he gave up his belief in Christianity. 'It is not supported by evidence', he concluded. Nevertheless, as far as we know, he never lost a belief in a supernatural creator behind nature.

Several types of evidence led Darwin to accept that species must evolve. On his voyage down the South American continent, he observed that related species gradually replaced one another. The species living in the Galapagos also puzzled him. Many of these were unique to the islands, yet most were strikingly similar to South American species. But the Galapagos, a collection of marine volcanoes, had never been connected to South America and their climate was totally different.

According to Lyell's view, species were somehow created to suit new environments. So why were the Galapagos species so obviously related to South American ones instead of just being rocky island species? Darwin's explanation was that their ancestors must have come from South America and changed over time.

In 1838, Darwin read Thomas Malthus's *An Essay on the Principle of Population* (1798), which argued that continued population growth would lead to famine and starvation. Darwin was struck by the implications of checks to population growth. This led him to focus on what allowed some individuals rather than others to survive and pass on their characteristics.

He hypothesized that every organism varied in many small ways, and any of these variations that helped or hindered would make a difference to which survived. He eventually called this filtering process **'natural selection'** by analogy with the process in which farmers changed domesticated plants and animals by selecting desirable individuals to breed from. In so doing, they emphasized some traits and reduced others (see 'Evolution in a nutshell').

FIGURE 1.2 Galapagos Island finches are the posterchild for Darwin's ideas about evolution, but their importance might have been overblown.

It would take Darwin more than 20 years to publish these ideas. By early 1858, he had drafted many chapters and was about a year or two away from publishing his 'big book', which would have spanned several volumes.

## Words from Wallace

Then, on 18 June, something surprising happened. An essay arrived in the mail from Alfred Russel Wallace, outlining a theory almost identical to Darwin's own.

Wallace was a brilliant collector who had worked in Southeast Asia since 1854. He had long been privately convinced that species must evolve. But he was certainly not, as many modern commentators put it, searching for a mechanism for how evolution works. As he collected thousands of tropical insects and birds, his theoretical views gradually matured.

In February 1858, on the tiny spice island of Ternate, Wallace lay sweating from fever when he thought of a means whereby species could become naturally adapted to a changing world. It was a filtering process of life and death that was very similar to Darwin's natural selection. When he recovered, Wallace wrote an essay, 'On the tendency of varieties to depart indefinitely from the original type', aimed primarily at Lyell's anti-evolutionary arguments. Shortly after, Wallace received an encouraging letter from Darwin stating that Wallace's hero, Lyell, admired

# 126,000
*Number of specimens Wallace collected on his Malay expedition*

Wallace's work. This led Wallace to send the essay to Darwin with the request that it be forwarded to Lyell.

Darwin was struck by the resemblance between Wallace's views and his own. The same day, Darwin sent the essay on to Lyell, bemoaning that he ought to send it for publication ahead of his own work. Ever the Victorian country gentleman (see 'Getting to know Darwin'), it seemed like the noble thing to do.

---

### Getting to know Darwin

For someone who devised a revolutionary idea, Charles Darwin lived a remarkably quiet life. In 1842, Darwin and his wife Emma moved from London to rural Kent in southern England. They already had two children then, and would go on to have eight more.

Darwin had very regular habits. He rose early and went for a walk. After breakfast he worked in his study until 9.30 am, his most productive time of the day, then read his letters lying on the sofa before returning to work.

At midday he would go for another walk accompanied by his dog, stopping at his greenhouse to inspect his botanical experiments. Then he would proceed to the sand walk, a gravel path around a strip of woodland. While strolling on this 'thinking path', Darwin would ruminate on his unsolved scientific problems.

After luncheon he read the newspaper and wrote letters. His network of correspondents provided information from all corners of the globe.

The Darwins were not very strict parents and the children were apt to run wild. Their mild-mannered father

---

worked patiently to a background of playful screams and little footsteps stampeding past his study door.

After dinner Darwin played backgammon with his wife. They were very competitive. He once wrote, 'Now the tally with my wife in backgammon stands thus: she, poor creature, has won only 2490 games, whilst I have won, hurrah, hurrah, 2795 games!'

Despite poor health, Darwin continued to publish a string of innovative and seminal works until his final book on earthworms in 1881. It was an instant bestseller. He died the following year, aged 73.

But science had one last claim on him. Rather than a quiet interment in the local churchyard, which he called 'the sweetest place on Earth', Darwin was given a state funeral in London's Westminster Abbey.

Lyell and another of Darwin's peers, Joseph Dalton Hooker, did not agree with that view. They had been aware of Darwin's theory for years and were not prepared to withhold their knowledge of Darwin's priority. They proposed a compromise: to present Wallace's essay together with some of Darwin's unpublished writings at a meeting of the Linnean Society of London. Modern opinions about these arrangements can be strong indeed, especially among those who think Wallace was unfairly treated. This is another mid-twentieth-century view. According to the standards of the time, however, the arrangement was fair. Wallace had sent his essay without asking for it to be kept private. The conventions of the day allowed Darwin or Lyell to publish it. Wallace was always accorded the honour of being the co-discoverer of natural selection and never tired of expressing his gratitude and satisfaction.

These brief writings of Darwin and Wallace offered the first statement of how species came into existence by natural means, yet they made remarkably little impact. Urged to bring out a reduced overview

**8 years**

*Length of Wallace's expedition to the Malay Archipelago*

of his massive work in progress, Darwin spent 13 months condensing his 20 years of study into a single volume. This was published on 24 November 1859 as *On the Origin of Species*.

---

**Evolution in a nutshell**

Darwin's and Wallace's theory of evolution maintains that new species are descended from earlier ones. This long-term process happens because all organisms vary. The tiny variations are naturally 'selected' by virtue of whether or not they help an organism to survive the brutal struggle for existence in nature. Many are born, but few survive; fortuitous variations are preferentially passed on. This process of endless filtering works to adapt organisms to their environment.

---

## On the Origin of Species

The book was immediately controversial and widely reviewed and discussed. Darwin came in for a great deal of ridicule and abuse. The implication that human beings must have evolved from earlier species was particularly objectionable to many, as was the revelation that no divine guiding hand was needed; species evolved on their own.

**20 years**

*Time taken for Darwin to work up and publish his ideas on natural selection*

But Darwin also gained strong support especially from members of the younger gen-eration of naturalists, such as Thomas Henry Huxley (today always referred to as 'Darwin's bulldog', but not known as this in his lifetime). Darwin's mass of evidence, ranging from embryology and vestigial organs to geographical distribution, and his arguments in favour of evolution were overwhelming.

Despite its baptism of fire, *On the Origin of Species* almost single-handedly convinced the international scientific com-munity that evolution was a fact. In his 1889 book *Darwinism*, Wallace wrote of the revolution Darwin effected: 'This totally unprecedented change in public opinion has been the result of the work of one man, and was brought about in the short space of twenty years!'

The theory of evolution has come a long way since. Today we think of it in terms of genes and DNA, but Darwin and Wallace had no idea of their existence. It was only in the 1930s and 1940s that genetics was incorporated into evo-lutionary theory (see Chapter 3). Even now, new discov-eries are shaking up our understanding, but at the core of the modern theory remains Darwin's idea of **descent with modification**.

## *Evolution vs creation: What really happened at the legendary 1860 debate?*

'The proprieties of the Association have been outraged.' So wrote civil servant Arthur Munby in his diary on 1 July 1860. And no wonder: the previous day, at the annual meeting of the British Association for the Advancement of Science in Oxford, there had been a most ungentlemanly discussion. The topic was Charles Darwin's new – and dangerous – idea. Bishop Samuel Wilberforce stood ready to defend the works of God in creation. The zoologist Thomas Huxley attended as evolution's champion. Much of what is widely believed about this legendary debate is far from reliable, written 20 years after the event by people with an axe to grind. Historian Frank James has spent ten years examining diaries, letters and other eyewitness accounts written within days or weeks of the event. These records, he says, lead us to a rather different conclusion about the outcome of that day.

It came as a last-minute call: the room was far too small and time was short. Only the library in the university's new museum could hold the anticipated crowd, and even this grand hall needed extra seating to accommodate the throng. A troupe of carpenters set to work, and the air filled with sawdust and the sound of frenetic hammering.

Why the rush? A dull American called John Draper had suddenly become the meeting's hot ticket. His subject, Darwin's evolutionary theory, was of wide and topical interest. *On the Origin of Species* had been published only seven months earlier, and emotions were running high. But Draper wasn't the attraction: word was that evolution's fiercest opponents, including the Bishop of Oxford, were planning to make a stand.

It could have been a one-sided debate: Darwin was not there – he was unwell – and the prime defender of evolution, Thomas Huxley, was not planning to attend. However, he relented the day before, when accused of deserting the cause.

## Monkey grandparents

The hall was packed. A few sceptics spoke first, including the president of the Royal Society. Then Wilberforce rose to great applause and proceeded to argue that humans must have been specially created, rather than evolving from non-human animals, because the idea was a central pillar of Christianity. At the end of his argument, according to the following week's edition of *The Press*, he asked the famous question: would Huxley prefer a monkey for his grandfather or grandmother?

Huxley's response was double-edged. Respectful, but not giving any ground, he referred to Wilberforce as an 'unscientific authority' but paid homage to the bishop's intellect. 'If I had to choose between being descended from an ape or from a man who would use his great powers of rhetoric to crush an argument, I should prefer the former,' he declared. According to *The Evening Star* of 2 July, Huxley then went on to defend Darwin's ideas 'in an argumentative speech which was loudly applauded'.

In his own version of events, written in a letter to the marine zoologist Frederick Dyster in September, Huxley painted himself in an even better light. There was, he wrote, 'inextinguishable laughter among the people' at his witty reply. 'I believe I was the most popular man in Oxford for full four-and-twenty hours afterwards.'

## Smashing Wilberforce

Others were not quite so impressed by Huxley's performance. Robert FitzRoy, captain of HMS *Beagle* during Darwin's all-important voyage, remained unconvinced by Huxley's arguments, and the archaeologist and biologist John Lubbock was left thinking Darwin's hypothesis was nothing more than the best on offer.

Indeed, when Joseph Hooker, assistant director of the Royal Botanic Gardens at Kew, related the day's events to Darwin, he grumbled that Huxley didn't 'put the matter in a form or way that carried the audience' so he'd had to do it himself. 'I smashed [Wilberforce] amid rounds of applause … Sam was shut up – had not one word to say in reply and the meeting was dissolved forthwith.'

Wilberforce didn't remember it that way. 'I think I thoroughly beat him,' he wrote to archaeologist Charles Anderson three days later. The physicist Balfour Stewart agreed. 'I think the Bishop had the best of it.'

Victory, it seems, lay in the eye of the beholder. After a decade delving through the documents, James, historian at London's Royal Institution, suggests that the popular view of Huxley's victory may have arisen only because Wilberforce was not well-liked, a fact missing from most accounts. 'Had Wilberforce not been so unpopular in Oxford he would have carried the day and not Huxley.'

But what does the official record say? Very little. The association's report for 1860 makes no reference to the discussion at all. 'The British Association had a gentlemanly ethos. And these were most ungentlemanly goings-on,' says James.

In the end, the gentlemen of the British Association suppressed as much information about the debate as possible, never realizing the significance of their decision. The story of Huxley's

FIGURE 1.3 It was while sweating with a fever on the spice island Ternate that Alfred Wallace devised his theory of evolution by natural selection.

victory only really took hold 20 years later – when it suited the cultural climate. 'In the 1880s there was a split developing between religion and science,' James says. 'In the 1860s there really wasn't one.' But, without any official record of the meeting to counter their claims, scientists looking to establish their authority were able to refer back to the debate as the moment when science defeated religion: the battle, they declared, was already won. 'You see it referred to as an enormously important event when, at the time, it quite clearly wasn't,' James says. But by turning this local discussion into a universal myth, men and women in the late nineteenth century contributed to the process of separating science from Christian belief.

## Interview: Writing *On the Origin* was 'like confessing a murder'

*150 years after the publication of* On the Origin of Species, New Scientist *magazine obtained an interview with its author.* *

**What was it like, coming up with the idea that changed the world?**

Like confessing a murder.

**The emotional and physical struggle you went through must have taken its toll?**

I have suffered from almost incessant vomiting for nine months, & that has so weakened my brain, that any excitement brings on whizzing & fainting feelings.

**You would clearly rather I didn't excite you further, but I must say that when I grasped your idea that life has been changing, evolving, for billions of years, I was captivated.**

You cannot imagine how pleased I am that the notion of Natural Selection has acted as a purgative on your bowels of immutability. Whenever naturalists can look at species changing as certain, what a magnificent field will be open.

**Quite so. Now I must put to you the question that authors are inevitably asked: how did you get your ideas?**

It seemed to me probable that allied species were descended from a common parent. But for some years I could not conceive how each form became so excellently adapted to its habits of life. I then began systematically to study domestic productions, & after a time saw clearly that man's selective power was the most important agent. I was prepared from having studied the habits of animals to appreciate the struggle for existence, & my work in geology gave me some idea of the lapse of past time. Therefore when I happened to read "Malthus on population" the idea of Natural selection flashed on me.

**The 'greatest idea anyone has ever had' just flashed on you! Your modesty and rigorous experimental approach are an inspiration to us all. But what would you say to young scientists starting out now?**

When I joined the "Beagle" as Naturalist I knew extremely little about Natural History, but I worked hard.

**You have been one of the most influential scientists of all time, yet your work continues to generate controversy, especially among those of a religious persuasion. You famously said there was grandeur in evolution, but does an atheist outlook help you get through the day?**

It has always appeared to me more satisfactory to look at the immense amount of pain & suffering in this world, as the inevitable result of the natural sequence of events, i.e. general laws, rather than from the direct intervention of God.

**Would you describe yourself as an atheist?**

In my most extreme fluctuations I have never been an atheist in the sense of denying the existence of a God. – I think that generally (& more and more so as I grow older) but not always, that an agnostic would be the most correct description of my state of mind.

**What about the perceived conflict between religious beliefs and your theories?**

It seems to me absurd to doubt that a man may be an ardent Theist & an evolutionist.

**Some of your bulldogs would deny that. They are sometimes accused of overdoing the fight against those who deny evolution.**

I feel sure that our good friend Huxley, though he has much influence, would have had far more if he had been more moderate & less frequent in his attacks.

**Your daughter Annie died when she was ten years old, and this tragic event is said to have influenced you greatly. Can you explain your feelings about this?**

Thank God she suffered hardly at all, & expired as tranquilly as a little angel. – Our only consolation is, that she passed a short, though joyous life. – She was my favourite child; her cordiality, openness, buoyant joyousness & strong affection made her most loveable. Poor dear little soul. Well it is all over.

**You are sometimes accused – unfairly, it seems to me – of racism. What are your views on slavery, which was still very widespread when you were on the *Beagle*?**

I have seen enough of Slavery & the dispositions of the negros, to be thoroughly disgusted with the lies and nonsense one hears on the subject in England... Great God how I shd [sic] like to see that greatest curse on Earth Slavery abolished.

**Have these views had an effect on your politics?**

I would not be a Tory, if it was merely on account of their cold hearts about that scandal to Christian Nations, Slavery.

**Thanks so much for agreeing to talk to us.**

I daresay you will say that I am an odious plague.

**Not at all. It has been an honour.**

*All of Darwin's quotes are taken from his vast correspondence, collected online at the Darwin Correspondence Project. The words attributed to Darwin are as he wrote them. The questions have been framed to conform as far as possible to the context in which Darwin was writing.

**What if Darwin had not sailed on the *Beagle*?**

Just how different would history be had Charles Darwin not gone on the five-year voyage around South America?

On at least one thing most Darwin scholars agree: had Darwin not sailed aboard the *Beagle* he would not have arrived at his theory of evolution by natural selection. It took the raw novelty of travel in alien climes to shatter his received views of nature as harmonious, benign and static. So many aspects of wild nature posed awkward, even embarrassing questions for this cosseted young man. And so, in the months after his return, he began questioning the unquestionable: the immutability of species.

Driven by what he had seen on the *Beagle* voyage, Darwin secretly became an evolutionist, and then in the winter of 1838 he came up with a plausible mechanism for this change: natural selection. It is hard to imagine Darwin making the same cognitive leaps if he were a vicar in rural England – the career path his father envisaged for him.

But does this really matter historically? After all, Alfred Russel Wallace arrived at pretty much the same theory. Perhaps, then, had Charles Darwin not sailed aboard the *Beagle* we would simply refer to 'Wallaceism' instead.

Possibly, but it would have been a hard fight. For a start, in 1858 Wallace had only a tiny fraction of the data available to Darwin. And Wallace's humbler origins would have made it harder for the establishment to accept the dangerous idea of evolution.

Without Darwin, we would still – mostly – believe in evolution by natural selection. To think otherwise would be to ignore the huge advances in biological science in the first half of the twentieth century that made the theory irresistible. But although we can't say how long it would have taken biologists to embrace evolutionism without him, one thing is fairly certain: in 1859 Charles Darwin gave the still-fragile theory of evolution the protection it needed to take root and become perhaps the most powerful idea of modern science.

## Route to a breakthrough

**1795**

Erasmus Darwin, Charles's grandfather, foreshadows Darwin when he writes: 'Would it be too bold to imagine that all warm-blooded animals have arisen from one living filament … with the power of acquiring new parts'.

**1798**

The doom-mongering 'An Essay on the Principle of Population' by Thomas Malthus warns of the dire consequences of unrestricted population growth. It is a key influence on Charles Darwin.

**1848**

Wallace departs for an expedition to Brazil. On the return journey four years later, a fire destroys many of his specimens.

**1837**

Darwin sketches the first 'tree of life' in his notebook to explain the evolutionary relationships between species.

**1858**

Wallace conceives of his theory of how species adapt to a changing world. On 1 July both Wallace's and Darwin's ideas are presented at the Linnean Society of London.

**1859**

Darwin's *On the Origin of Species* is published and becomes the object of much ridicule and abuse.

**1809**
French naturalist Jean-Baptiste Lamarck
publishes *Philosophie Zoologique*,
outlining his idea of evolution according
to a 'complexifying force'.

**1809**
Charles Robert Darwin is
born in Shrewsbury, UK,
the fifth of six children
in a prosperous family.

**1831**
Darwin sets sail on
HMS *Beagle* for a
survey of South
America. The voyage
lasts five years.

**1823**
Alfred Russel Wallace
is born in the village of
Llanbadoc near the
border of England and Wales,
the seventh of nine children.

**1813**
French zoologist Georges
Cuvier publishes 'Essay
on the Theory of the Earth',
setting out his idea that
new species appear after
catastrophes such as floods.

**1860**
Debating evolution at Oxford, Bishop
Samuel Wilberforce asks Thomas
Huxley, a champion of Darwin's ideas,
if Huxley's monkey ancestors were on
his grandmother's or grandfather's side.

**1870**
Evolution comes to be
accepted as fact by most
of the international
scientific community.

# 2

# But what exactly is evolution?

*Evolution is the unifying force in modern biology; it ties together fields as disparate as genetics, microbiology and palaeontology. It is an elegant and convincing explanation for the staggering diversity of Earth's nine million or so living species. Here is a primer to the basics.*

Evolution has several facets. The first is the theory that all living species are the modified descendants of earlier species, and that we all share a common ancestor in the distant past. All species are therefore related via a vast tree of life. The second is that this evolution is driven by a process of natural selection – or the 'survival of the fittest'.

Darwin argued that all individuals struggle to survive on limited resources, but some have small, heritable differences that give them a greater chance of surviving or reproducing, than individuals lacking these beneficial traits. Such individuals have a higher evolutionary fitness, and the useful traits they possess become more common in the population because more of their offspring survive.

Eventually these advantageous traits become the norm. Conversely, harmful traits are quickly eradicated as individuals that possess them are less likely to reproduce. Natural selection therefore works to create a population that is highly suited to its environment, and can adapt to changes.

## Sex wars

When individuals compete for limited resources in their environment they are subject to ecological selection. However, useful traits are not only those that give a survival advantage, but also those that increase a plant or animal's chance of reproducing. These traits are subject to sexual selection.

Sexually selected traits can make a male organism more attractive to females: the peacock's tail for example. These are sometimes correlated to the health of an individual, and are therefore an honest badge of fitness. Another type of sexually selected trait gives males a physical advantage in out-competing other males for mates: the stag's antlers for example. Sexual selection can even act at a molecular level.

FIGURE 2.1 The peacock is the poster child for the concept of sexual selection. Characteristics such as the peacock's tail evolve because individuals possessing them acquire more mates than their competitors.

Birds are particularly known for showy ornaments that attract mates, but which also increase the chances of being spotted by predators. Other sexually selected traits include: lions' manes, the plumage of great tits or budgies, grouse mating rituals, insect love tokens, the height of human males and human hair, intelligence and facial features. But natural and sexual selection are not the only factors that give rise to biological change. Random **genetic drift** is another factor. This could be called 'survival of the luckiest' (see Figure 6.2).

## Species spawning

Although Darwin titled his book *On the Origin of Species*, speciation was one thing he could not explain. He called it the 'mystery of mysteries', and even a century and a half later the mechanism by which two groups of animals become genetically incompatible remains one of the greatest puzzles in biology.

We understand how Darwin's Galapagos finches could have evolved from a single species – different populations became isolated and gradually adapted to different environments until they were no longer able to reproduce with each other.

This 'allopatric speciation' happens when a geographical change – a river changing course for example or a new mountain range – splits a species in two. Once separated, as happened to antelope squirrels on either side of the Grand Canyon in the US, the populations evolve independently, eventually becoming distinct and reproductively isolated.

However, speciation also occurs rapidly and without physical isolation of populations, which is far harder to explain, and researchers are still trying to pinpoint the exact biological mechanisms that underly it. Examples of this 'sympatric speciation' include the 13 species of Galapagos finch or Africa's cichlid fish. These species adapt to different opportunities in the environment, and then cease to interbreed – perhaps due to some isolating mechanism. New species can also form through hybridization, such as sunflowers. (See 'How new species are formed' in Chapter 7 for more on this)

Like individuals in a population, species also struggle to survive, and most become extinct over time. Species can also die out in mass extinctions, such as the one that caused the demise of the dinosaurs. Today we may be in the throes of another mass extinction, caused by human over-exploitation of habitats.

## Evolutionary scenarios

During his voyage on the HMS *Beagle* and throughout his life, Darwin gathered evidence that contributed to his theory of natural selection. In *On the Origin of Species* he presented support from the fields of embryology, geography, palaeontology and comparative anatomy. Darwin also found evidence for his theory in examples of convergent evolution, co-evolution and adaptive radiation.

**Convergent evolution**, is when the same adaptations have evolved independently in different lineages of species under similar selection pressures. Today we see convergent evolution in species as diverse as: shark and camels, shrimps and grasshoppers, flamingos and spoonbills, marsupial and placental mammals and bioluminescent sea creatures. We also see it in the ears and teeth of mammals.

**Co-evolution** is when the evolutionary history of two species or groups of species is intimately intertwined. Examples include: the co-evolution of flowering plants and pollinators such as bees, lizards and moths; pocket gophers and their lice; humans and intestinal microbes; and the war our immune systems wage with the pathogens that attack us.

**Adaptive radiation** is the rapid speciation of one ancestral species to fill many empty ecological niches. Adaptive radiations are most common when animals and plants arrive at previously uninhabited islands. Examples of adaptive radiation can be found in: the Galapagos finches, Australia's marsupials, Hawaii's honeycreepers and fruit flies, Madagascar's carnivores and other mammals, New Zealand's birds and the prehistoric flying pterosaurs.

## How did the giraffe get its long neck?

Most people assume that giraffes' long necks evolved to help them feed. If you have a long neck, runs the argument, you can eat leaves on tall trees that your rivals can't reach. But there is another possibility: the prodigious necks may have little to do with food, and everything to do with sex.

The evidence supporting the high-feeding theory is ambiguous. Giraffes in South Africa do spend a lot of time browsing for food high up in trees, but research from Kenya showed that they don't seem to bother, even when food is scarce.

Long necks come at a cost. Because a giraffe's brain is around 2 metres above its heart, the heart has to be big and powerful. In fact, for the blood to reach the brain it has to be pumped at the highest pressure of any animal. So there must be a big payback to keep giraffes' necks so long.

A competing theory is that the long necks are the result of sexual selection: that is, they evolved in males as a way of competing for females.

Male giraffes fight for females by 'necking'. They stand side by side and swing the backs of their heads into each others' ribs and legs. To help with this, their skulls are unusually thick and they have horn-like growths called ossicones on the tops of their heads. Their heads, in short, are battering rams. Having a long and powerful neck would be an advantage in these duels, and it's been found that males with long necks tend to win, and also that females prefer them.

The 'necks for sex' idea is contentious, but also helps explain why giraffes have extended their necks so much more than their legs. If giraffes evolved to reach higher branches, we might expect their legs to have lengthened as fast as their necks, but they haven't.

The problem for the sex idea is that it implies that female giraffes shouldn't have long necks, and they plainly do. One suggestion is that giraffes' necks may have begun growing as a way of eating hard-to-reach food, but that they were then 'hijacked' for mating purposes. Once the necks had reached a certain length, males could use them for necking and clubbing – and at that point sexual selection took over, driving the necks to their current extreme lengths.

## Secret code

Darwin was able to establish natural selection, without any understanding of the genetic mechanisms of inheritance, or the source of novel variation in a population. His own theory on the transmission of traits, called pangenesis, was completely wrong.

It was not until the start of the twentieth century that the genetic mechanism of inheritance began to be revealed (see Chapter 3 for more on this).

### Interview: Wonders of nature have been my solace in life
*Alfred Russel Wallace (1823–1913) discovered the theory of evolution by natural selection independently of Charles Darwin and founded the science of evolutionary biogeography. His correspondence – from which the answers for this 'interview' are mined – is now available online: he tells us of his research, expeditions and enduring fascination for nature's mysteries.*

**You are famously joint author, with Darwin, of the first paper describing the origin of species and natural selection, published in 1858. When did you first get the idea?**

I began [in 1847] to feel rather dissatisfied with a mere local collection – little is to be learnt by it. I sh[ould]d like to take some one family, to study thoroughly – principally with a view to the theory of the origin of species. By that means I am strongly of [the] opinion that some definite results might be arrived at.

**This desire led you to Brazil to collect birds, butterflies and beetles to try to discover what drives the evolution of new species. Were there any incidents on the voyage?**

On Friday the 6th of August [1852]... the Captain (who was the owner of the vessel) came into the cabin & said "I am afraid the ship's on fire. Come & see what you think of it."

**Despite that harrowing experience, you next undertook an eight-year expedition to the Malay Archipelago, where you discovered the invisible boundary between the animals of Asia and the Australian region, which would later be called the Wallace Line in your honour. What fascinated you most on that trip?**

The Birds have however interested me much more than the insects, they are proportionally much more numerous, and throw great light on the laws of Geographical distribution of Animals in the East... As an instance I may mention the Cockatoos, a group of birds confined to Australia & the Moluccas, but quite unknown in Java, Borneo,

Sumatra & Malacca… Many other species illustrate the same fact.

**You have been famously good-natured about sharing the discovery of natural selection with Darwin…**

I also look upon it as a most fortunate circumstance that I had a short time ago commenced a correspondence with Mr. Darwin on the subject of "Varieties", since it has led to the earlier publication of a portion of his researches & has secured to him a claim to priority which an independent publication either by myself or some other party might have injuriously effected.

**What did you and Darwin have in common?**

In early life both Darwin and myself became ardent beetle-hunters. Both Darwin and myself had, what he terms "the mere passion of collecting"… Now it is this superficial and almost child-like interest in the outward forms of living things, which, though often despised as unscientific, happened to be *the only one* which would lead us towards a solution of the problem of species.

**Do you feel your contribution has been overlooked?**

The idea came to me, as it had come to Darwin, in a sudden flash of insight: it was thought out in a few hours – was written down with such a sketch of its various applications and developments … then copied on thin letter-paper and sent off to Darwin – all within one week.

I should have had no cause for complaint if the respective shares of Darwin and myself in regard to the elucidation of nature's method of organic development had been thenceforth estimated as being, roughly, proportional to the time we had each bestowed upon it when it was thus first given to the world – that is to say, as 20 years is to one week.

**How did you feel looking back on your life's work, at the age of 89?**

The wonders of nature have been the delight and solace of ... life. Nature has afforded ... an ever increasing rapture, and the attempt to solve some of her myriad problems an ever-growing sense of mystery and awe.

# 3
# Darwin and DNA: How genetics spurred the evolution of a theory

*Darwin and Wallace's theory of evolution famously lacked a mechanism. Then, in the early twentieth century, came the new field of genetics. Although genetics revolutionized ideas about inheritance, at first it was not obvious that it had anything to do with evolution. So how did DNA become incorporated in our modern understanding of evolution?*

## *The genetic revolution of evolution*

Our understanding of evolution today stems from the combination of two very different ideas. One came from a monk who studied pea plants in a Moravian monastery in the 1850s. The other came from Darwin and Wallace.

Although Gregor Mendel and Charles Darwin were alive at the same time, they never met and Darwin wasn't aware of Mendel's work. With hindsight, the union of the two men's work seems like a marriage made in heaven (or hell, if you're a creationist). In fact, for many years, it wasn't obvious that Mendel's studies of heredity had any relevance to Darwin's theory of evolution by natural selection. It would take nearly 60 years for this jigsaw to be pieced together and give rise to the 'modern synthesis' of evolution, which framed Darwin's idea in terms of genetics.

How exactly did this new understanding arise? And why did it take so long?

The explanation starts with natural selection itself. According to this, only the fittest – the best adapted to the local environment – survive and breed, and in this way the population as a whole gradually transforms. The idea of evolution was already accepted by many biologists in the mid-nineteenth century, but there was considerable opposition to the notion that it happened by means of natural selection.

The plausibility of this mechanism rests on the assumption that beneficial characteristics are passed more or less intact from one generation to the next. But it was not clear how this might happen.

## Explaining inheritance

To explain heredity, Darwin proposed a hypothesis he called pangenesis. It posited that each organism produces particles called 'gemmules', which transmit its characteristics to the next generation. Darwin suggested that the offspring develops from a mix of the parents' gemmules and thus exhibits a blend of their characters.

But the idea had a major flaw, seized upon by his opponents: blending would result in the useful characteristics of one parent becoming diluted as it mated with individuals that do not have those traits. Over successive generations these characteristics would gradually disappear. It was a problem no one was able to solve during Darwin's lifetime.

Unbeknown to Darwin and his compatriots, the key to solving this puzzle had already been found. Sometime in the 1840s, Gregor Mendel joined the friary at Brno, in what is now the Czech Republic. In the years that followed, he made detailed studies of how the characteristics of pea plants were passed from one generation to the next. He found that parents' traits were not blended in their offspring; rather, they were transmitted unchanged in predictable ratios. This led him to devise laws of inheritance, published in 1866 (see 'Who was Gregor Mendel?').

No one imagined, though, that the characteristics Mendel studied in peas, such as flower colour, were of any general significance, and the work was largely ignored for decades.

---

### Who was Gregor Mendel?

Gregor Mendel had an unlikely background for someone known as the founder of modern genetics, not least because he carried out his work 50 years before genes were actually discovered. Born in 1822 on a farm in what is now the Czech Republic, he joined the Augustinian friary at Brno and here carried out his work on heredity.

In the friary garden he bred thousands of pea plants, noting the presence of characteristics such as flower colour and wrinkliness of seeds. He found, for example, that when he crossed white-flowered plants with purple ones, the resulting plants weren't a light mauve, as might be expected if parental traits blended together, but appeared as either white or purple in fixed ratios.

These observations led him to devise the now famous laws of inheritance, published in 1866, which introduced the idea of dominant and recessive traits. This work went largely unnoticed until the turn of the century, when his ideas were incorporated into the new science of genetics.

The significance of Mendel's contribution is debated, however. His laws certainly helped to clarify how characteristics are transmitted from parent to offspring. But even Ronald Fisher, who used Mendel's ideas to create a genetic theory of evolution in the 1930s, suggested that Mendel's results were a bit too good to be true, perhaps 'tidied up' by an overzealous assistant. And it is by no means clear that Mendel would have been a proponent of the theory that was based on his work. He expressed his ideas solely in terms of the transmission of characters from one generation to the next – there was no discussion of any mechanism.

Mendel eventually abandoned his studies when he became abbot at the age of 46. Little else is known about him as his correspondence and other personal papers were burned after his death.

FIGURE 3.1 As a monk, Gregor Mendel had an unlikely background
for the founder of modern genetics.

Then in 1900, Mendel's laws were rediscovered by the
botanists Hugo DeVries and Carl Correns. Through studying inheritance, each independently came to the view
that an organism's characteristics are fixed units that are
transmitted unchanged. Only later did they discover that
Mendel had carried out similar work.

## Dawn of the gene

A new science of heredity emerged. First dubbed 'Mendelism', it was soon christened 'genetics' by the biologist William Bateson, who translated Mendel's paper into English and was a key promoter of his work. Bateson derived the name from the Ancient Greek word *genesis*, meaning 'origin'.

Mendel had expressed his laws in terms of characteristics transmitted from parent to offspring. The early geneticists were convinced that some material entity in the organism must encode that information.

Before long, the biologist Thomas Hunt Morgan had identified genes as units arranged along the chromosomes inside the cell's nucleus. Working on the fruit fly *Drosophila* in 1910, Morgan showed that the trait for eye colour could be traced to a specific place on the X chromosome. This led to a burst of discoveries about the links between different genes, and to the creation of genetic maps showing the positions of genes on chromosomes.

Morgan's research eventually won him a Nobel prize and confirmed that genes were the physical substance of inheritance. It would take another three decades, however, to discover that they were made of **DNA** and that each **gene** codes for a specific protein.

The concept of the gene seemed to be the missing piece in Darwin's jigsaw. It completed his picture of natural selection by showing that traits can't be blended away to insignificance, although this wasn't recognized immediately.

Genetics also solved another problem of Darwin's theory: the source of the variation within a population. Darwin's starting point was that any population naturally contains a variety of individuals, providing the raw material for natural selection. A key source of this variation was now shown to be mutation – spontaneous changes in the structure of a gene, leading it to code for something new. Such changes had been observed by Morgan and others as they traced the position of the genes on chromosomes.

Morgan himself came to believe that harmful mutations would quickly be eliminated from a population, thus recognizing the negative side of natural selection. However, more work was needed to demonstrate how selection acted on genes to create positive evolutionary change.

## Slow or quick?

According to Darwin, evolution is a slow process of gradual **adaptation** to the environment, in which most characteristics have, or once had, an adaptive function. So giraffes with slightly longer necks were able to reach leaves higher up, and thus gradually evolved longer necks through the process of natural selection. In contrast, many of the original geneticists saw evolution as something that happened in large jumps, or saltations, whereby new characteristics appear abruptly as the result of some internal rearrangement of an organism's hereditary constitution. For example, a plant could suddenly start producing flowers of a colour not seen in its parents. The change would not necessarily have any adaptive benefit.

The early geneticists were attracted to Mendel's laws precisely because they seemed to support these ideas. Morgan thought that 'Nature makes new species outright' through a 'sudden change of the germ'. Bateson saw no value in the Darwinians' studies of continuous variation and resisted the claim that natural characteristics appear as a result of adaptive pressures on the species. By the same token, the saltation mechanism of change seemed to have no relevance to the process of natural selection.

These entrenched positions made it hard for anyone to suggest a way to reconcile the two approaches. But that changed in the 1920s, thanks to the new field of population genetics – the study of how certain genes within populations change over time.

The biologists Ronald Fisher, J.B.S. Haldane and Sewall Wright used sophisticated mathematical models to show that natural selection is able to enhance the frequency of any gene coding for a beneficial character, and eliminate those that are maladaptive.

This concept was developed in Fisher's 1930 book *The Genetical Theory of Natural Selection* and in Haldane's more popular *The Causes of Evolution* in 1932. That same year, Wright introduced the idea of an adaptive landscape, a map depicting all possible gene combinations and the resultant fitness of the organism.

---

### From genetics to eugenics

Some of the early pioneers of evolutionary theory were enthusiastic proponents of eugenics: the idea of enhancing the human population by eliminating 'unfit' genes.

Ronald Fisher, for example, devoted part of his 1930 book *The Genetical Theory of Natural Selection* to his hopes for improving the human race by this means. He even fathered eight children to further the cause.

The dark reality of eugenics became clear in the early twentieth century when several US states legislated to sterilize the 'feeble-minded', and the Nazis took the idea to its ultimate, horrific extreme.

---

Collectively, their work demonstrated that genes accounted for both the abrupt changes in characteristics sometimes seen in an organism's offspring, and the continuous variation that

Darwin had documented for large populations. These biologists showed that genetic selection is a genuinely creative force driving the adaptation of a species to its local environment, with continual mutation ensuring that the fund of variability is maintained. However, their theoretical models involved complex statistics and were hard to understand.

The gene-centred perspective of evolution only reached the wider scientific community in 1937, when Theodosius Dobzhansky published his *Genetics and the Origin of Species*, translating the mathematical formulations into terms that were more accessible. Dobzhansky's work also expanded our understanding of how genetics enabled evolution, showing, for example, how new species could emerge when isolated populations changed to adapt to their local environment.

In 1942, the biologist Julian Huxley's broad survey *Evolution:The Modern Synthesis* gave the new perspective a name. By the 1950s, this formulation had become dominant, although one key aspect would continue to be debated for decades to come.

## A higher purpose?

More or less since its conception, Darwinian evolution was seen as an idea hostile to the Christian vision of nature as the product of some higher purpose. In the US especially, creationist opposition to Darwinism took hold in the 1920s and has continued ever since.

The founders of the **modern synthesis** wanted to present Darwinism as able to accommodate the belief that evolution has a built-in tendency to produce higher levels of organization. Dobzhansky, for example, came from a Russian Orthodox background and wrote *Mankind Evolving* in 1962 to promote the idea that evolution had an ultimate purpose. Huxley also wrote prolifically to promote the idea of evolutionary progress. These authors presented ideas about the modern synthesis in a way that did not challenge traditional hopes and values too openly. This, however, did not stop some of them being proponents of eugenics (see 'From genetics to eugenics', above).

The quasi-religious portrayal would change in later decades, most notably with the publication of Richard Dawkins' *The Selfish Gene* in 1976 and his emergence as a leading proponent of the argument that nature has no ultimate moral purpose. Subsequent debates over the evolution of social behaviour and the emergence of altruism have taken place against the backdrop of an increasing tension between Darwinian evolution and religion – exactly what the founders of the synthesis hoped to avoid.

Despite these difficulties, the modern synthesis remains at the heart of our understanding of evolution today. It is itself evolving as advances in genetics, developmental biology and ecology broaden our understanding of the relationship between genes, organisms and the environment.

The gene-centred view of evolution that emerged from the ideas of Darwin and Mendel is being transformed by the growing recognition that the environment in which the organism develops does play a role in shaping its characteristics, and may even affect the way traits are passed on to future generations. Discoveries in the field of epigenetics are showing that chemical tags that attach to genes to switch them on and off might be as important for development as the hard-wired genetic code itself (see Chapter 8).

The modern synthesis was an idea for the twentieth century. In the twenty-first century, the story of evolution is set to acquire a sophistication that Darwin could only have dreamed of (see Chapter 11).

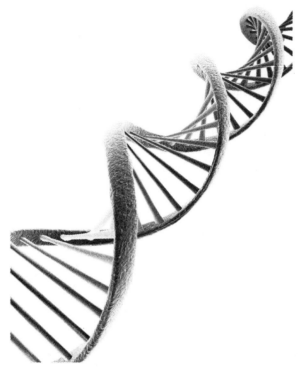

FIGURE 3.2 The double helical structure of DNA discovered in 1953.

## What exactly is a gene?

In general, a gene consists of the DNA sequence that codes for a protein, along with regulatory sequences, such as a promotor, that help determine when, where and how much of the protein is made. In complex cells, the coding sequence is split into several parts, called exons, separated by longer bits of junk DNA called introns.

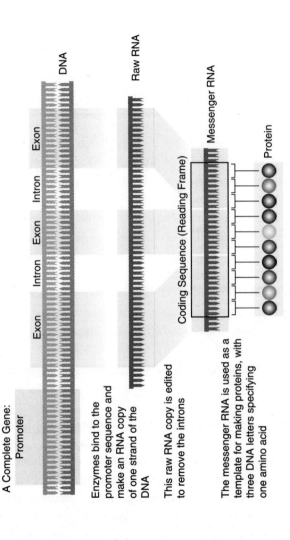

FIGURE 3.3  What is a gene?

49

## *How genes evolve*

As the genomes of more and more species are sequenced, we can not only trace how the bodies of animals have evolved, we can even identify the genetic mutations behind these changes.

Most intriguing of all, we can now see how genes – which are the recipes for making proteins, the building blocks of life – arise in the first place. And the story is not unfolding quite as expected.

The most obvious way for a new gene to evolve is through the gradual accumulation of small, beneficial mutations. Less obvious is how an existing gene that already does something important can evolve into a different gene. The scope for such a gene to change tack without capsizing the organism that carries it is very limited. However, as biologists realized a century ago, this constraint no longer applies when mutations produce an entire extra copy of a gene.

According to the textbooks, the process by which new genes form starts with gene duplication. In the vast majority of cases one of the copies will acquire harmful mutations and will be lost. Just occasionally, though, a mutation will allow a duplicate gene to do something novel. This copy will become specialized for its new role, while the original gene carries on performing the same task as before.

Surprisingly, gene duplication has turned out to be nearly as common as mutations that change a single 'letter' of DNA code. During the exchange of material between chromosomes prior to sexual reproduction, mistakes can create extra copies of long DNA sequences containing anything from one gene to hundreds. Entire chromosomes can be duplicated, as happens in Down's syndrome, and sometimes even entire genomes.

Since duplication can throw up trillions of copies for evolution to work with, it is not surprising that over hundreds of millions of years, a single original gene can give rise to many hundreds of new ones. We humans have around 400 genes for smell receptors alone, all of which derive from just two in a fish that lived around 450 million years ago.

## Not the whole story

This classical view of gene evolution is far from the whole story, however. Genes often have more than one function, so what happens when a gene is duplicated? If a mutation knocks out one of the two functions in one of the copies, an organism can cope fine because the other copy is still intact. Even if another mutation in this other copy knocks out the second function, the organism can carry on as normal. Instead of having one gene with two functions, the organism will now have two genes with one function each – a mechanism dubbed subfunctionalism. This process can provide the raw material for further evolution.

But the real challenge to the classical model comes from actual studies of new genes in various organisms. A comparison of the genomes of several closely related species of fruit fly, for example, identified new genes that have evolved in the 13 million years or so since these species split from a common ancestor. This revealed that around 10 per cent of the new genes had arisen through a process called retroposition. This occurs when messenger RNA copies of genes – the blueprints sent to a cell's protein-making factories (see Figure 3.2) – are turned back into DNA that is then inserted somewhere else in the genome. Many viruses and genetic parasites copy themselves through retroposition, and the enzymes they produce sometimes accidentally retropose the RNA of their host cells.

This process may have created many of the recently evolved genes in us apes. A burst of retroposition in our ancestors, peaking around 45 million years ago, gave rise to many thousands of gene duplicates, of which at least 60 or 70 evolved into new genes. The burst was probably due to a new genetic parasite invading our genome.

The evolution of new genes often involves even more drastic changes. In fruit flies, for example, a third of new genes were significantly different from their parent genes, having lost parts of their sequences or acquired new stretches of DNA.

Where do these extra sequences come from? In complex cells, the DNA coding for a protein is broken into several parts, separated by non-coding sequences. After an RNA copy of the entire gene is made, the non-coding bits – the introns – are cut out and the coding parts – called exons – are spliced together. This edited RNA copy is then sent to a cell's protein-making factory. The modular form of genes greatly increases the chances of mutations reshuffling existing genes and generating novel proteins. There are all sorts of ways in which it can happen: exons within a gene can be lost, duplicated or even combined with exons from different genes to create a new, chimeric gene.

## Variations on a theme

For instance, most monkeys produce a protein called TRIM5, which protects them from infection by retroviruses. In one macaque in Asia around 10 million years ago, an inactive copy of a gene called CypA, produced by retroposition, was inserted near the TRIM5 gene. Further mutation resulted in cells producing a chimeric protein that was part TRIM5, part CypA.

This protein provides better protection against some viruses. Although it might seem an unlikely series of events, in fact the TRIM5-CypA gene has evolved not once but twice – much the same thing happened in owl monkeys in South America.

Given enough time – or rather enough mutations – gene duplication and reshuffling can produce new genes that are very different from the ancestral ones. But are all new genes variations on a theme, or can evolution throw up new genes unlike any that already exist?

A couple of decades ago, it was suggested that unique genes could arise from what is called a frameshift mutation. Each **amino acid** in a protein is specified by three DNA 'letters', or bases – the triplet codon. If a mutation shifts the starting point for reading codons – the 'reading frame' – by one base, or by two, the resulting protein sequence will be completely different. Since DNA is double-stranded, any given piece can be 'read' in six different ways.

## Gibberish

The vast majority of mutations that alter the reading frame of a gene produce nonsense, usually dangerous nonsense. Many genetic diseases are the result of frameshift mutations wrecking proteins. It's a bit like swapping every letter for the next one along in the alphabet: the result is usually gibberish. But not always.

Another source of unique new genes could be the 'junk' DNA littering most genomes. An early hint this might be so came two decades ago when a team at the University of Illinois revealed the genesis of the antifreeze protein produced by one Antarctic fish. The gene involved originally coded for a digestive enzyme. Then, around 10 million years ago, as the world's climate cooled, part of one of the introns – a piece of junk DNA, in other words – got turned into an exon and subsequently duplicated many times, generating the characteristic repetitive structure of antifreeze proteins. From a random bit of DNA evolved a gene vital to the fish's survival.

Still, the antifreeze gene evolved from a pre-existing gene. What are the chances of mutations in junk DNA generating an entire new gene from scratch? Practically zero, most biologists thought until very recently. After all, it takes a whole set of unlikely conditions for a piece of random DNA to evolve into a gene. First, some of the DNA must act as a promoter, telling the cell to make RNA copies of the rest. Next, these RNA copies must have a sequence that can be edited into a viable messenger RNA blueprint for the protein-making factories. What's more, this messenger RNA must encode a relatively long protein – the average length is 300 amino acids – which is unlikely because in a random stretch of DNA, on average 1 in 20 every codons will be a 'stop' codon. Finally, of course, the new protein must do something useful. The obstacles seemed insurmountable.

This view changed in 2006, when David Begun of the University of California, Davis, and colleagues identified several new genes in fruit flies with sequences unlike any of the older genes. They suggested that these genes, which code for relatively small proteins, have evolved from junk DNA in the past few million years. A few years later, a hunt for new genes in fruit flies revealed another nine genes that appear to have evolved from scratch out of junk DNA. Another study found evidence that at least six new human genes have arisen from non-coding DNA since humans and chimps diverged more than 6 million years ago.

How can the number be so high when the likelihood of a gene evolving in this way is so rare? Part of the answer could be the recent discovery that even though at least half of our genome is junk, as much as 90 per cent of it can be accidentally transcribed into RNA on occasion.

This means it might not be that uncommon for random bits of junk DNA to get translated into a protein. Since most random proteins will probably be harmful, natural selection will eliminate these DNA sequences, but just occasionally one will strike it lucky. A sequence that does something beneficial will spread through a population and rapidly evolve into a new gene, becoming optimized for whatever role it plays.

It will be many years yet before we have a clear picture of the relative importance of the various mechanisms by which genes can evolve. What is certain, though, is that the classical view of how they evolve is far from complete. Evolution isn't fussy – it'll take new genes wherever it can get them.

And as sequence data continues to pour in, biologists are well on their way to working out how every one of our 20,000 or so genes evolved.

---

### The Selfish Gene

Richard Dawkins' 1976 book *The Selfish Gene* popularized the notion that the gene, rather than the individual, was the true unit of evolution. In it he wrote that people are 'robot vehicles blindly programmed to preserve the selfish molecules known as genes'. The notion of the 'selfish gene' has dominated evolutionary genetics ever since and is the most successful scientific metaphor of recent times, followed not far behind by 'the extended phenotype'. Both terms were coined by Richard Dawkins (see interview in Chapter 9) and are the titles of his first popular science books.

---

The message in the *The Selfish Gene* is that evolution is about the natural selection of genes, and genes alone. Dawkins sees them as the best candidates to be evolution's units of replication. As such, the genes that are passed on are those whose consequences serve their own interests at gene level – that is, to continue being replicated – and do not necessarily serve the interests of the organism at a larger level, or at the level of groups of organisms.

Dawkins' *The Extended Phenotype* (1982) develops this idea, arguing that in their drive for survival and replication, genes extend their influence beyond the appearance, or phenotype, of an individual and into the world where it also affects their chance of survival. Think of the beaver's dam or spider's web. However, many biologists think it's time to rethink this gene-centred view of evolution (see Chapter 11).

## *Five classic examples of gene evolution*

As the genomes of more and more species are sequenced, geneticists are piecing together an extraordinarily detailed picture of the molecules that are fundamental to life on Earth. With modern techniques, we can not only trace how the bodies of animals have evolved, we can even identify the genetic mutations behind these changes and, as we recently reported, genes sometimes evolve in surprising ways. Here though, in celebration of the versatility of DNA, are five classic examples of gene evolution.

## Collecting colours

Ever noticed how dogs sometimes seem unable to spot a brightly coloured ball that's obvious to you? That's because most mammals have just two colour-sensitive retina pigments, or opsins, compared to our three. This means they effectively have a form of colour blindness.

So why do we have three? In the ancestors of apes and some monkeys, the gene MWS/LWS – which codes for one of the two pigments found in most mammal eyes – was duplicated. Spare gene copies usually degenerate quickly as they acquire mutations, but in this case mutations in one copy resulted in an opsin that could detect a different spectrum of light. In this way, we gained better, trichromatic colour vision.

There is, however, a twist to this tale. The ancestors of vertebrates actually had even better colour vision than we do, thanks to their four colour-sensitive opsins. Unlike us, they could see ultraviolet, as well as the other colours. This ability has been inherited by most amphibians, reptiles and birds, so how come mammals lost two of the colour-sensitive opsin genes?

The answer is probably that because some of the earliest mammals were nocturnal, they had little need for colour-sensitive opsins that only worked during daylight. As a result, these genes acquired mutations and some were lost – if you don't use it, you lose it.

Our vision could have evolved along very different lines. When the ancestors of geckos became nocturnal, they evolved colour night vision.

## Crystal clear

You wouldn't be able to read these words without the crystallin proteins in your eye. These transparent proteins can bend light thanks to their high refractive index, enabling the lens of the eye to focus light on the retina. So where did evolution find transparent proteins with a high refractive index when the eye was evolving? All over the place, it turns out.

Take alpha-crystallin, which is found in a variety of animal eyes, including ours. It was originally a heat-shock protein – a type of protein that keeps other proteins in shape. In fact, it still is a heat-shock protein. In some tissues in the body, where only small amounts of the protein is made, it still carries out this role – in the lens, though, it is produced at high levels and its main function is optical.

There is only one gene, HspB5 that codes for alpha-crystallin. So the evolution of a new function – such as bending light – does not always require the evolution of an entire new gene encoding a novel protein. Sometimes, all it takes are a few mutations in the sequences that determine how much of an existing protein is made in a particular tissue type. Sometimes evolution takes the easy path.

---

### Memes: Evolution doesn't just happen to genes

The term 'meme' was coined by biologist Richard Dawkins in his 1976 book, *The Selfish Gene*, which explored the principles of Darwinism. His idea was that Darwin's principle of evolution by natural selection need not apply just to biology. Given some kind of copying machinery that makes lots of slightly different copies of the same information, and given that only a few of those copies survive to be copied again, an evolutionary process must occur.

---

The information that is copied, varied and selected is called the replicator, and the process is well understood when applied to biology. In biological evolution, the replicators are genes, but there is no reason why there should not be other evolutionary systems, with other replicators. So Dawkins invented the term 'meme' as a cultural replicator.

Everything you have learnt by copying it from someone else is a meme. This includes your habit of driving on the left or right, eating beans on toast, wearing jeans or going on holiday. You would do none of these things if someone else hadn't done them, or something very like them, before you did. Imitation, unlike other forms of learning, is a kind of copying or replication. Other animals can be masters of learning, as when squirrels remember their hundreds of food stores, or cats and dogs build extensive mental maps. But this is learning by association, or trial and error. Only by imitation are the fruits of the learning passed on from one animal to the next – and humans are unrivalled when it comes to copying one another.

The idea of memes as a replicator in their own right has been much maligned and many biologists reject it. Yet memetics has much to offer in explaining human nature. According to meme theory, humans are radically different from all other species because we alone are meme machines. Human intelligence is not just a bit more or a bit better than other kinds of intelligence, it is something completely different, based on a new evolutionary process and a new kind of information.

## Fishy smell

Over hundreds of millions of years, a single gene can give rise not just to one new gene, but to hundreds via gene duplication. We humans have around 400 genes coding for smell receptors, for instance, all of which derive from two original genes in a very early fish living around 450 million years ago.

The evolution of this gene 'family' has been a messy process. Genome studies show that, rather than steadily acquiring genes for new smell receptors, there have also been extensive losses of genes during the evolution of mammals – a process dubbed 'birth-and-death evolution'.

This has led to great variation between mammals. You would expect dogs to have more receptors than humans, with around 800 working smell-receptor genes – but why do cows have even more, with over 1000?

Molecular evolutionary biologist Masatoshi Nei has suggested that mammals only need a certain minimum number of different olfactory receptors to have a good sense of smell. What animals do with the ones they have – the wiring of the brain during development, in other words – may matter more when it comes to a keen sense of smell.

As long as animals have more olfactory receptor genes than they need, he suggests, there is no natural selection and genes are gained and lost randomly. In other words, genetic drift might explain why the numbers and type of smell receptors vary so widely among mammals.

## Double for nothing

The HOX genes are a family of closely related genes that control embryonic development in animals. They are the 'master switches', the proteins that coordinate the activation of other sets of genes during development.

All HOX genes evolved from a 'protoHOX' gene in a very early animal. This protoHOX gene was repeatedly duplicated, creating a cluster of 13 HOX genes in the ancestor of vertebrates. Then the entire genome in this ancestral line got duplicated. And then it got duplicated again, creating the four clusters of HOX genes that control the development of all living vertebrates.

In the lineage leading to mammals, 13 of the 52 genes created by genome duplications were lost, leaving mammals with 39 HOX genes. The real mystery, though, is why so many of the gene copies created by the genome duplications survived? Why didn't they just degenerate and disappear? It might make sense to keep spare copies of genes handy, but evolution does not plan for the future.

The same phenomenon has been seen in the African clawed frog, *Xenopuslaevis*, whose entire genome got duplicated 40 million years ago. The vast majority of all these extra gene copies should have croaked long ago. Yet even after all this time, up to half of the duplicate genes have been preserved.

For instance, in a remarkable 2006 study Mario Capecchi, of the Howard Hughes Medical Institute in Salt Lake City, reversed the process that gave rise to the HOX gene family. He merged two existing HOX genes, HOXA1 and HOXB1, to recreate the HOX1 gene from which they evolved. Mice given this ancestral gene instead of the two modern ones still developed normally.

His work suggests that the two new genes together do no more than the ancestral gene did. In other words, both the gene copies did degenerate after being duplicated. No advantage has been gained by swapping one gene for two: the process was neutral.

This phenomenon, proposed in 1999, is known as subfunctionalization: when a gene is duplicated, the functions of the original gene can end up divided among the copies. Studies of the clawed frog suggest subfunctionalization can explain the preservation of at least a third of gene copies.

What this shows is that greater genomic complexity – having more genes – can evolve as a result of genetic drift, as well as by natural selection. Once organisms have acquired extra genes, of course, there is more chance of those genes acquiring beneficial new functions that will be selected for.

### Enigmatic enzyme

Nylon was first made in 1935. Just 40 years later, in 1975, a bacterium was discovered that is able to digest and live off not nylon itself, but waste chemicals from its manufacture – chemicals that had not existed before nylon production began.

It was later shown this bacterium, now known as Arthrobacter KI72, has evolved several types of enzymes capable of utilizing these waste products. One type, 6-aminohexanoic acid hydrolase, encoded by genes called nylBs, has become known popularly as 'nylonase'.

As a dramatic example of evolution in action, nylonase has attracted a lot of attention over the years. But there has also been a great deal of confusion about how it evolved.

In 1984, geneticist Susumu Ohno suggested that one way in which new genes could evolve is through a 'frameshift' mutation – one that alters the way in which the genetic code is read and thus completely alters the amino acid sequence of a protein. And nylonase evolved this way, he claimed.

Then in 1992, another team claimed that nylB genes are unique and had evolved by a rather complicated and special mechanism.

They are both wrong, says Seiji Negoro of the University of Hyogo, Japan, whose team has published many studies on the structure and evolution of nylon-related enzymes. His team's study of the protein structure shows that nylonase is very similar to a common type of enzyme that breaks down beta-lactamases – natural antibiotics produced by many organisms. Just two amino-acid changes – two mutations, in other words – are required to change the beta-lactamase binding site to one capable of binding the nylon by-product.

However, while Ohno was wrong about nylonase, he was right about frameshift mutations being one way in which genes evolve. Hundreds of examples have now been discovered in humans alone.

---

## Who needs new genes?

To do new things or make new body parts, organisms don't necessarily need to evolve entire new genes. Identical proteins often take on different roles in different parts of the body, while a single gene can produce many proteins.

Alternative splicing of RNAs – including some parts of a gene but not others – can generate a huge variety of different proteins. Studies show that alternative splicing is far more common in people than was thought, with most genes producing at least two variants. One human gene, bn2, can generate more than 2000 different proteins, some of which have no similarity at all. The fruit fly gene, Dscam, can produce an astonishing 38,000 variants.

That's not all. RNAs from two different genes can be edited together to generate a new protein, in a process called trans-splicing, which can greatly increase the number of potential proteins.

---

# Route to a gene-centred view of evolution

**5000 BC**
Humans begin to understand inheritance when they start to selectively breed more useful varieties of livestock and crops such as maize, wheat and rice.

**400 BC**
Ancient Greek philosophers contemplate mechanisms of human inheritance. Hippocrates believed that the material of heredity was tiny particles in the body which accumulate in a seminal fluid in the parents. These particles blend to create the traits of the offspring.

**1937**
Theodosius Dobzhansky develops the modern synthesis, defining evolution in genetic terms as the 'change in the frequency of an allele [gene type] within a gene pool'.

**1920s**
The new field of population biology begins to unite the ideas of Darwin and Mendel, establishing how evolution can work at the level of genes.

**1942**
Ernst Mayr outlines how new species can evolve, for example when a geographical barrier results in a population becoming genetically incompatible with its original species.

**1944**
DNA is proven to be the material of heredity, not protein as had been suspected.

## 1859

Charles Darwin publishes *On the Origin of Species* – his explanation of evolution by natural selection. It contains a wealth of evidence for how variable traits become more common in a population, but suggests no mechanism for their transmission.

## 1866

Augustinian monk Gregor Mendel publishes meticulous studies of inheritance in pea plants, marking the birth of modern genetics. The findings go unnoticed for over three decades.

## 1905–6

The term 'genetics' is coined by the biologist William Bateson, a key proponent of Mendel's work. Soon the concept of a gene is developed.

## 1900

Mendel's laws of inheritance are rediscovered by Dutch and German botanists.

## 1868

Darwin publishes *The Variation of Animals and Plants under Domestication*, outlining his hypothesis of pangenesis: that particles called gemmules transmit an organism's characteristics to its offspring.

## 1951

Images of DNA are captured for the first time by Rosalind Franklin. Two years later, James Watson and Francis Crick determine the double-helix structure of DNA.

## 1990

The Human Genome Project begins, completing its work 13 years later when the full sequence is revealed. Genomes of many other organisms follow.

# 4
# How life began

*Evolution needs something to work on, but for many millions of years after it formed, Earth was a lifeless place with hellish conditions. Then around 3.8 billion years ago, after the surface had cooled and oceans formed, something amazing happened. Out of Earth's primordial chemicals arose an entity capable of replicating itself. Life was born. But how exactly did life get kick-started?*

## Meet your maker

What was the last shared ancestor of all life like? And how did it make its living? We may never know for sure, but researchers are beginning to home in on this long-lost life form.

The only thing we know for certain is that life must have popped into existence sometime between Earth's formation 4.5 billion years ago and the appearance of the first undisputed fossils, about 3.4 billion years ago.

In 1859, when Darwin published *On the Origin of Species*, he dedicated an entire chapter to the problem of missing 'intermediate links' – transitional forms that bridged the evolutionary gaps between closely related species. If his theory was correct, the fossil record should be full of them. Where were they?

At the time it was a real problem, as few such fossils had been found. Then came the spectacular discovery, in 1861, of Archaeopteryx, with the wings and feathers of a bird and the teeth and tail of a dinosaur.

Since then we have discovered a multitude of intermediate links: fish that could crawl, lizards with mammal-like jaws, whales with legs, giraffes with short necks and many others. But there's one we are unlikely ever to find: the link between the earliest proto-life and life as we know it, also known as the last universal common ancestor, or LUCA.

LUCA lived around 4 billion years ago – a tiny, fragile life form that is the direct ancestor of every single living thing, from aardvarks to zebras. It wasn't the very first life: thousands, if not millions, of years of evolutionary experimentation preceded it. But understanding LUCA would give us our best view yet of the origin of life.

We already know a surprising amount. Although any traces LUCA left in rocks were probably obliterated aeons ago, we can find clues to their nature in every living cell today, including our own. Cells use the same genetic code embodied in DNA. That would suggest that the ancestor of all living things, LUCA, was made of DNA.

## Which came first?

At first, that idea threw up a chicken-and-egg problem. All life uses proteins to carry out its essential functions, including making DNA and executing its code. But proteins themselves are made from DNA templates – so without DNA, no proteins. Which came first?

The solution appears to be neither. RNA is a close relative of DNA. It, too, is found in all living cells, and also carries the genetic code. Unlike DNA, RNA molecules come with their own toolbox, acting as enzymes and catalysing chemical reactions. Our current best theory of life's origins, known as the RNA world hypothesis, says the genetic code was born out of an early soup of RNA molecules that eventually gave rise to DNA and the first cells. Indeed, not all life made the switch – some viruses are still RNA-based.

But this still leaves a problem: if LUCA was made of RNA, where did RNA come from? Darwin himself was among the first scientists to speculate on how life originated: he envisaged a 'warm little pond, with all sorts of ammonia and phosphoric salts, lights, heat, electricity, etc present'. In the 1950s, US chemists Stanley Miller and Harold Urey famously tried to find out by zapping a mixture of gases and water with electricity. They ended up with a handful of biotic molecules – and naturally concluded that with a bit of extra va-va-voom, dead stuff could turn into live building blocks.

FIGURE 4.1 Was life's kitchen a warm muddy pond?

Nowadays, though, this idea of a bolt from the blue has been overtaken by more nuanced ideas. Nick Lane at University College London, for example, thinks that warm vents on the ocean floor – 'black smokers' with their soup of methane, minerals and water – would have provided the right conditions to form RNA. Michael Yarus of the University of Colorado in Boulder, meanwhile, favours the idea of a slushy pond. The continual freezing and thawing could have pushed chemicals together in just the right way, he says.

Or maybe it was some combination of both, maybe neither. Intriguingly, more recent experiments trying to coax RNA into existence have shown that when the chemistry is just right, many of the building blocks seem to form all by themselves. That widens the possibilities for how life started. Not just that: if the chemistry of life came naturally to our planet, why not elsewhere, too?

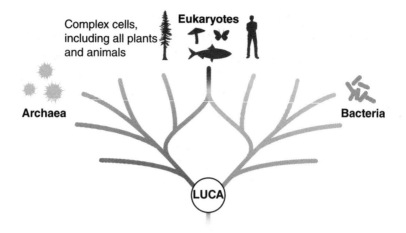

FIGURE 4.2 Meet your maker: We're getting closer to understanding what the last universal common ancestor of all life on Earth (LUCA) was like and where it came from.

## Oldest traces of life on Earth

The hunt for the oldest known fossils is a contentious area. Fossilized microbes are hard to distinguish from mineral structures that have nothing to do with life, and the same goes for the geochemical traces of ancient life.

The oldest compelling fossil evidence for cellular life was found on a 3.43-billion-year-old beach in western Australia in 2011. Grains of sand there provided a home for cells that dined on sulphur in a largely oxygen-free world.

The rounded, elongated and hollow tubular cells – probably bacteria – were found to have clumped together, formed chains and coated sand grains. Similar sulphur-processing bacteria are alive today, forming stagnant black layers beneath the surface of sandy beaches.

The remarkably well-preserved three-dimensional fossil microbes were excavated from an ancient beach – now a sandstone formation. Grains of unoxidized iron pyrite were found among the microbes, showing that there was no oxygen present at the time.

Since this discovery, claims of even older fossils have been reported – but the jury is still out as to whether these structures are evidence of life or have a non-biological origin.

In 2016 the journal *Nature* published a report of structures discovered in 3.7 billion-year-old rocks in Greenland that appeared to be evidence of microbes living in a shallow sea on early Earth. The structures, no more than a few centimetres tall, look like stromatolites (layered mounds that were – and still are – formed by photosynthetic microbes living in water). Analysis showed that the structures have the same chemical signature as sea water, pointing to a marine origin. To seal the case, it would have been ideal to have found fossil remains of microbes within the rock, but this is not possible with this type of rock.

## Even older?

Then in 2017 came an even more remarkable – and controversial – discovery. Matthew Dodd at University College London, and colleagues, analysed rocks collected from a region called the Nuvvuagittuq belt, in northern Quebec, Canada. The rocks here are at least 3.75 billion years old, and some geologists argue they are about 4.29 billion years old, which would mean they are just slightly younger than the planet itself.

Like all such ancient rocks, they have been heavily altered. At some point they spent time deep inside Earth, where temperatures above 500°C and extreme pressures baked and deformed them. But geologists can still read clues that they formed at the bottom of Earth's very early oceans. Significantly, they seem to preserve evidence of ancient deep-sea hydrothermal vents – just the sort of environment that many see as the most likely birthplace for life.

Dodd and his colleagues believe they have found evidence of early life, in iron-rich rocks originally formed around relatively cool vents (less than 160°C). These rocks contain microscopic tubes and filaments made of iron oxide. Similar structures are formed by bacteria that live in mat-like colonies around modern deep-sea hydrothermal vents.

What's more, the material close to the filaments includes carbon with an isotopic balance characteristic of biological processes. Some of that carbon is inside crystals of phosphorus-rich minerals, which also hints at early biology, as phosphorus is essential for all life on Earth.

Taken together, the evidence points to one inescapable conclusion, says Dodd: the very early Earth was home to microbes similar to those found around today's low-temperature hydrothermal vents.

If confirmed, the conclusion would be significant for several reasons. It would potentially push the record of life back to 4.29 billion years ago, suggesting our planet was inhabited astonishingly early.

More than that, it would show that life got going around deep-sea vents where organisms have to derive their energy from geothermal processes, since there is little or no sunlight. This would help bring the geological evidence in line with findings from genetic and biochemical studies hinting that life emerged in deep hydrothermal areas – and not in shallow, sun-drenched environments where most early fossils have been found.

But not everyone is so sure that these structures are evidence of life, or that fragile, microscopic structures could survive in rocks that have been subjected to high temperatures and pressures deep underground.

It may be many years before a consensus is reached. When dealing with 'fossils' that may be 4.29 billion years old, it makes little sense to rush things.

## *Life on Earth may have emerged not once, but many times*

In 4.5 billion years of Earthly history, life as we know it arose just once. Every living thing on our planet shares the same chemistry, and can be traced back to LUCA, the last universal common ancestor. So we assume that life must have been really hard to get going, only arising when a nigh-on-impossible set of circumstances combined.

Or was it? Simple experiments by biologists aiming to recreate life's earliest moments are challenging that assumption. Life, it seems, is a matter of basic chemistry – no magic required, no rare ingredients, no bolt from the blue.

And that suggests an even more intriguing possibility. Rather than springing into existence just once in some chemically blessed primordial pond, life may have had many origins. It could have got going over and over again in many different forms for hundreds of thousands of years, only becoming what we see today when everything else was wiped out in Earth's first ever mass extinction. In its earliest days on the planet, life as we know it might not have been alone.

Just to be clear, what we are talking about came long before animals or plants or even microbes. We are going right back to the start, when the only things fitting the description of 'life' were little more than molecular machines. Even then, having stripped away bodies, organs and cells and reduced everything down to the essential reactions, things appear devilishly complex. At a bare minimum, life needs some kind of code, it needs to use that code to make useful molecular machines, and then the code must be able to make copies of itself.

Over the decades, people have invoked all sorts of external forces to explain how some of the starting components were made. In the famed Urey–Miller experiments of the 1950s, the trigger was a zap of electricity mimicking a lightning bolt striking water (see 'Did life need a bolt from the blue to get kick-started?'). Other theories have invoked extraterrestrial delivery by meteorites or comets.

---

### Did life need a bolt from the blue to get kick-started?

In the 1950s, two chemists – Stanley Miller and Harold Urey – were the first to show that some of the essential building blocks for life can be made from simpler materials. The critical step was electricity. They mixed water with gases they thought would have been present on early Earth and zapped them with simulated lightning. This produced amino acids, the molecules that all modern proteins are made of.

Life did not necessarily need proteins to get started, but they certainly became necessary at some stage, and all living things that exist today rely on the same 20 or so amino acids to make proteins. It is becoming clear that amino acids must form very easily indeed: they have been found almost anywhere in space astrobiologists care to look. Some have been found on meteorites and by the Rosetta probe orbiting comet 67P/Churyumov-Gerasimenko.

Experiments still suggest there must have been some input of energy to produce these amino acids – be it shock waves from a meteorite impact or heat transmitted from deep in the Earth via a hydrothermal vent. So although building a genetic code may not have required a bolt from the blue, building proteins probably did.

---

FIGURE 4.3 Life might have needed a bolt from the blue to get kick-started.

## Where did it begin?

More recently, chemists interested in the origins of life have taken a more methodical approach to the problem. By breaking down the very beginning into its component stages (see Figure 4.2), they are stripping away the mystique that surrounds that initial spark of life. What they have discovered points to a very different beginning.

For Philipp Holliger of the MRC Laboratory of Molecular Biology in Cambridge, UK, the difference between life and non-life is genetic code. 'Biology has memory, while chemistry does not,' he says. 'To me, the origin of life is really the origin of information.' Many biologists subscribe to the RNA world hypothesis for the beginnings of life, which says that before DNA, this information was embodied in its close relative, RNA. Both molecules are long strings made of repeating units, or 'letters'. So step one to building life, in this version of events, has to be making the building blocks of RNA.

In May 2016, Thomas Carell, a biomolecular chemist at the Ludwig Maximilian University of Munich, Germany, announced that his team had found a very easy way to make some of these units, from substances that could have been abundant on early Earth: hydrogen cyanide, ammonia and formic acid. His reactions hint that an RNA world could have been made relatively easily. 'You really don't need special conditions,' he says. 'These reactions can happen everywhere – a little pond, the deep sea, wherever.'

So we have the letters of the code, though they're not much use in free-floating form. The next stage might be easy, however. Twenty years ago, British chemist Leslie Orgel showed that if you could only get the building blocks of RNA to form they would spontaneously assemble into chains. All he needed was clay. That's because crystals in some clays carry a natural electrical charge which appears to pull RNA letters in and encourage them to line up and stick to each other.

## From code to machines

We are still some way from life, though. Code is only useful if it acts as a template to make things like proteins – the building materials and engines of all living things. In our bodies, this is known as gene expression, and is an incredibly complicated process overseen by sophisticated molecular machines. They are very unlikely to have emerged just so from the primordial mud. How do you do away with all that in the earliest life forms? Michael Yarus at the University of Colorado in Boulder

believes he has the solution. His team found a surprisingly simple reaction which they say looks like very rudimentary gene
expression. By mixing repeating strands of RNA in water with
extra free-floating RNA letters, they found that the letters

① Make genetic building blocks

② Assemble them into chains to create
a template or code

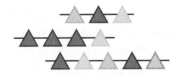

③ Use the code to make rudimentary
molecular machines

④ Replicate the code to make new
living things

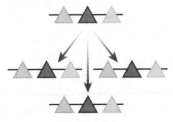

FIGURE 4.4 The four steps to creating life.

would spontaneously arrange themselves to form new molecules. The original strand of RNA seemed to act as a template. Intriguingly, the new molecules look like some of the simplest chemical machines inside our own bodies, called coenzymes.

But other steps are needed to create life. Ask anyone what they think the key properties of life are and sooner or later most will say 'reproduction'. Living things make copies of themselves, inert things like rocks do not. Without reproduction, life is a dead end.

Today, we have special enzymes tasked with replicating DNA. But in June 2016 a team lead by Jack Szostak, an evolutionary chemist at Harvard Medical School, showed that RNA can efficiently copy itself without help from any enzymes. They mixed an RNA template with free-floating RNA building blocks just as Yarus did. But Szostak added a few RNA fragments that matched parts of the template. And that made all the difference. The fragments seemed to kick-start a replication process and soon the team had reasonably faithful copies of the templates.

'These reactions happen pretty easily,' says Szostak. Others have been able to copy RNA without enzymes before, but these reactions are faster. He says the small booster fragments are so short they could have formed spontaneously 4 billion years ago.

Szostak's team think these reactions or similar ones could have been an early form of replication, even though they might not have generated perfect copies each time. Eventually, new templates would have evolved, coding for really useful inventions like building cell walls. At that point, better copying would have become important.

## Multiple beginnings

Taken together, all these findings suggest that building a rudimentary RNA world may not have been the special, once-in-a-universe occurrence it is popularly made out to be. This raises an intriguing possibility: that life's earliest stages didn't happen just once, but over and over again.

If this is true, then life's first epoch was one of great experimentation. Many different kinds of live molecular machines would have popped up in the primordial soup, some more successful than others. For a time, they would have coexisted, but eventually only the most successful remained, either because it was better than everything else, conditions changed and favoured it, or by sheer chance. This would have been Earth's very first mass extinction, billions of years ago.

---

### The unlikely evolution of complex life – and life on other planets

It is generally assumed that once simple life has emerged, it gradually evolves into more complex forms, given the right conditions. But that's not what happened on Earth. After simple cells first appeared, there was an extraordinarily long delay – nearly half the lifetime of the planet – before complex ones evolved. What's more, simple cells gave rise to complex ones just once in 4 billion years of evolution: a shockingly rare anomaly, suggestive of a freak accident.

If simple cells had slowly evolved into more complex ones over billions of years, all kinds of intermediate cells would have existed and some still should. But there are none. Instead, there is a great gulf. On the one hand, there are the **prokaryotes** (**bacteria** and **archaea**), tiny in both their cell volume and genome size. On the other, there are the vast and unwieldy eukaryotic cells. A typical single-celled **eukaryote** is about 15,000 times larger than a bacterium, with a genome to match.

---

All the complex life on Earth – animals, plants, fungi and so on – are eukaryotes, and they all evolved from the same ancestor. So without the one-off event that produced the ancestor of eukaryotic cells, there would have been no plants and fish, no dinosaurs and apes. Simple cells just don't have the right cellular architecture to evolve into more complex forms, according to biologist Nick Lane of University College London.

To grow larger and more complex, he argues, cells have to generate more energy. And the way eukaryotes got around this problem was by acquiring other bacteria which evolved into tiny power generators called mitochondria. These provided the energy needed to create complex life. 'It's hard to imagine any other way of getting around the energy problem – and we know it happened just once on Earth because all eukaryotes descend from a common ancestor,' he says.

The emergence of complex life, then, seems to hinge on a single fluke event – the acquisition of one simple cell by another. Lane also claims that the extraordinarily low probability of this freak event explains why we haven't found evidence of alien life elsewhere in the universe.

## Horizontal gene transfer: Uprooting the tree of life

In July 1837, Charles Darwin had a flash of inspiration. In his study at his house in London, he turned to a new page in his red leather notebook and wrote, 'I think'. Then he drew a spindly sketch of a tree.

As far as we know, this was the first time Darwin toyed with the concept of a 'tree of life' to explain the evolutionary relationships between different species. It was to prove a fruitful idea: by the time he published *On the Origin of Species* 22 years later, Darwin's spindly tree had grown into a mighty oak. The book contains numerous references to the tree and its only diagram is of a branching structure showing how one species can evolve into many.

The tree concept has been the unifying principle for understanding the history of life on Earth. Darwin assumed that descent was exclusively 'vertical', with organisms passing traits down to their offspring. But what if species also routinely swapped genetic material with other species, or hybridized with them? Then that neat branching pattern

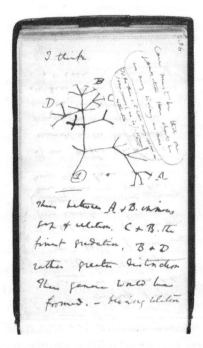

FIGURE 4.5 Darwin's first sketch of an evolutionary tree of life.

would quickly degenerate into an impenetrable thicket of interrelatedness, with species being closely related in some respects but not others.

We now know that this is exactly what happens, mainly in the realm of single-celled organisms. With the advent of advanced gene sequencing technology, it has become clear that the patterns of relatedness of two of the major domains of life – bacteria and archaea, collectively known as prokaryotes – could only be explained if they were routinely swapping genetic material with other species, often across huge taxonomic distances in a process called horizontal gene transfer (HGT).

Typically around 10 per cent of the genes in many bacterial genomes seem to have been acquired from other organisms in this way, though the proportion can be several times that. So an individual microbe may have access to the genes found in the entire microbial population around it, including those of other microbe species.

Surprisingly, HGT also turns out to be common in the third great domain of life, the eukaryotes. For a start, it is increasingly accepted that the eukaryotes originated by the fusion of two prokaryotes, one bacterial and the other archaeal, forming this part of the tree into a ring rather than a branch.

The neat picture of a branching tree is further blurred by a process called endosymbiosis. Early on in their evolution, eukaryotes are thought to have engulfed two free-living prokaryotes. One of these gave rise to the cellular power generators called mitochondria while the other was the precursor of the chloroplasts, in which photosynthesis takes place. These 'endosymbionts' later transferred large chunks of their genomes into those of their eukaryote hosts, creating hybrid genomes.

Other cases of HGT in multicellular organisms are coming in thick and fast (see 'Blurring the animal family tree'). The

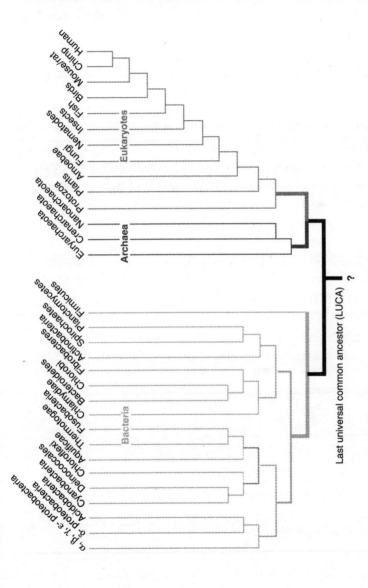

FIGURE 4.6 A simplified version of the tree of life, showing relationships between groups that had had their genomes sequenced. Constructing the tree has been a major aim of biology, but some now consider the enterprise to be misconceived in the light of current knowledge

human genome might be also harbouring a surprising number of genes gleaned from other organisms. A 2015 study pinpointed 145 genes in human DNA that seem to have jumped from simpler organisms.

---

### Blurring the animal family tree

There are many examples of animals acquiring genes 'horizontally' from bacteria, viruses and even other animals.

- The cow genome contains a piece of snake DNA that appears to have entered horizontally around 50 million years ago.
- The human gene *syncytin*, which is vital for placenta formation, originated in a virus.
- The tiny eight-legged tardigrades, renowned for their extreme survival skills, scavenged up to one-sixth of their DNA (and many of their protective genes) from bacteria and other organisms.
- The entire genome of the bacterium *Wolbachia* was found integrated into the genome of a fruit fly. The fly is, in effect, a fly-bacterium chimera.

---

The number of horizontal transfers in animals is not as high as in microbes, but can be evolutionary significant. Even so, nobody is arguing – yet – that the tree concept has outlived its usefulness in animals and plants. While vertical descent is no longer the only game in town, it is still the best way of explaining how multicellular organisms are related to one another. In that respect, Darwin's vision has triumphed: he knew nothing of micro-organisms and built his theory on the plants and animals he could see around him.

But it is clear that the Darwinian tree is no longer an adequate description of how evolution in general works. Some evolutionary relationships are tree-like, many others are not.

---

### Tree? Bush? Why does it matter?

Hang on, you may be thinking. Microbes might be swapping genes left, right and centre, what does that matter? Surely the stuff we care about – animals and plants – can still be accurately represented by a tree, so what's the problem?

For a start, biology is the science of life, and the first life was unicellular. Microbes have been living on Earth for at least 3.8 billion years; multicellular organisms didn't appear until about 900 million years ago. Even today bacteria, archaea and unicellular eukaryotes make up at least 90 per cent of all known species, and by sheer weight of numbers almost all of the living things on Earth are microbes. It would be perverse to claim that the evolution of life on Earth resembles a tree just because multicellular life evolved that way. 'If there is a tree of life, it's a small anomalous structure growing out of the web of life,' says John Dupré, a philosopher of biology at the University of Exeter, UK.

---

### *The evolution of the genetic code*

How did evolution produce the genetic code and the basic genetic machinery used by all organisms? Most biologists, following Francis Crick (co-discoverer of the structure of DNA) supposed that these were 'accidents of history'. But pioneering microbiologist Carl Woese (now deceased) and physicist Nigel Goldenfeld took a close look at this early stage of life on Earth and came to a startling conclusion: that Darwinian evolution simply cannot explain how such a code could arise. But horizontal gene transfer can.

Though the genetic code was discovered in the 1960s, no one had been able to explain how evolution could have made it so exquisitely tuned to resisting errors. Mutations happen in DNA coding all the time, and yet the proteins it produces often remain unaffected by these glitches.

The essence of the genetic code is that sequences of three consecutive bases, known as codons, correspond to specific amino acids. Proteins are made of chains of amino acids, so when a gene is transcribed into a protein these codons are what determines which amino acid gets added to the chain. The codon AAU represents the amino acid asparagine, for example, and UGU represents cysteine. There are 64 codons in total and 20 amino acids, which means that the code has some redundancy, with multiple codons specifying the same amino acid.

## The perfect code

This code is universal, shared by all organisms, and biologists have long known that it has remarkable properties. In the early 1960s, for example, Woese himself pointed out that one reason for the code's deep tolerance for errors was that similar codons specify either the same amino acid or two with similar chemical properties. Hence, a mutation of a single base, while changing a codon, will tend to have little effect on the properties of the protein being produced.

In 1991, geneticists David Haig and Lawrence Hurst at the University of Oxford went further, showing that the code's level of error tolerance is truly remarkable. They studied the error tolerance of an enormous number of hypothetical genetic codes, all built from the same base pairs but with codons associated randomly with amino acids. They found that the actual code is around one in a million in terms of how good it is at error mitigation. That would seem to demand some evolutionary explanation. Yet, until now, no one has found one. The reason, say Woese and Goldenfeld, is that everyone has been thinking in terms of the wrong kind of evolution.

Working with biologist Kalin Vetsigian, Woese and Goldenfeld set up a virtual world in which they could rerun history multiple times and test the evolution of the genetic code under different conditions. Starting with a random initial population of codes being used by different organisms – all using the same DNA bases but with different associations of codons and amino acids – they first explored how the code might evolve in ordinary Darwinian evolution. While the ability of the code to withstand errors improves with time, they found that the results were inconsistent with the pattern we actually see in two ways. First, the code never became shared among all organisms – a number of distinct codes remained in use no matter how long the team ran their simulations. Second, in none of their runs did any of the codes evolve to reach the optimal structure of the actual code.

# Timeline: The evolution of life

*Pinning down when specific events occurred is often tricky, and depends on the dating of rocks in which fossils are found, and by looking at the 'molecular clocks' in the DNA of living organisms. But there are difficulties with both these methods, meaning that the dates in the timeline should be taken as approximate. As a general rule, they become more uncertain the further back along the geological timescale we look.*

**3.8 billion years ago**
This is our best guess for the beginning of life on Earth. At some point after this, a common ancestor gave rise to two main groups of life: bacteria and archaea.

**3.4 billion years ago**
The first photosynthetic bacteria evolve.

**540 million years ago**
The Cambrian explosion begins, with many new body layouts appearing on the scene.

**630 million years ago**
Some animals evolve bilateral symmetry for the first time: that is, they now have a defined top and bottom, front and back. The first bilateral animal was a kind of worm.

**530 million years ago**
The first true vertebrate – an animal with a backbone – appears. It probably resembles an eel-like fish such as a lamprey or hagfish.

**500 million years ago**
Fossil evidence shows that animals were exploring the land at this time.

**150 million years ago**
Archaeopteryx, the famous 'first bird', lives in Europe.

**200 million years ago**
Proto-mammals evolve warm-bloodedness – the ability to maintain their internal temperature, regardless of the external conditions.

**75 million years ago**
The ancestors of modern primates split from the ancestors of modern rodents and lagomorphs (rabbits, hares and pikas). The rodents go on to be astonishingly successful, eventually making up around 40 per cent of modern mammal species.

**65 million years ago**
The Cretaceous-Tertiary (K/T) extinction wipes out a swathe of species, including most dinosaurs. The extinction clears the way for the mammals, which go on to dominate the planet.

**2.1 billion years ago**
Eukaryotic cells – cells with internal 'organs' – come into being.

**1.5 billion years ago**
The eukaryotes divide into three groups: the ancestors of modern plants, fungi and animals split into separate lineages.

**800 million years ago**
The early multicellular animals undergo their first splits. First they divide into, essentially, the sponges and everything else.

**900 million years ago**
The first multicellular life develops.

**489 million years ago**
The Great Ordovician Biodiversification Event begins, leading to a great increase in diversity. Within each of the major groups of animals and plants, many new varieties appear.

**400 million years ago**
The oldest known insect lives around this time. Some plants evolve woody stems.

**250 million years ago**
The greatest mass extinction in Earth's history occurs, wiping out swathes of species. Afterwards, the sauropsids (the group that includes modern reptiles, plus the dinosaurs and birds) dominate. The ancestors of mammals survive as small, nocturnal creatures.

**397 million years ago**
The first four-legged animals, or tetrapods, evolve. They conquer the land, and give rise to all amphibians, reptiles, birds and mammals.

**63 million years ago**
The primates split into two groups. One group becomes the modern lemurs and aye-ayes. The other develops into monkeys and apes – and humans.

**6 million years ago**
Humans diverge from their closest relatives; the chimpanzees and bonobos. Shortly afterwards, hominins begin walking on two legs.

## Horizontal is optimal

The results were very different when they allowed horizontal gene transfer between different organisms. Now, with advantageous genetic innovations able to flow horizontally across the entire system the code readily discovered the overall optimal structure and, came to be universal among all organisms. For the researchers the conclusion is inescapable: the genetic code must have arisen in an earlier evolutionary phase dominated by horizontal gene transfer.

Pinning down the details of that early process remains a difficult task. However, the simulations suggest that horizontal gene transfer allowed life in general to acquire a unified genetic machinery, thereby making the sharing of innovations easier. Hence, the researchers suspect that early evolution proceeded through a series of stages before the Darwinian form emerged, with the first stage leading to the emergence of a universal genetic code. Following this, a second stage of evolution would have involved rampant horizontal gene transfer, made possible by the shared genetic machinery, and leading to a rapid, exponential rise in the complexity of organisms. This, in turn, would eventually have given way to a third stage of evolution in which genetic transfer of the core functions of the cell became mostly vertical. This new Darwinian epoch came about because horizontal gene transfer of these core functions ceased be to effective beyond a certain time, as there was nothing new to transfer.

It seems that life arose from a collective, networked phase where there was no notion of species and perhaps even individuality itself.

## Exploding biodiversity

Three billion years after it first formed, life suddenly exploded into variety. The first of these bursts of animal diversification is known as the Cambrian explosion, which got started around 540 million years ago. In the space of just 20 million years at the start of the Cambrian, all bar one of the basic types, or phyla, of animals we see around us today, make their appearance in the fossil record.

The second great burst in animal evolution is known as the Great Ordovician Biodiversification event (see Figure 4.7). It began some 489 million years ago, when massive algal blooms provided a bountiful food supply which fuelled an even bigger evolutionary bonanza than the Cambrian.

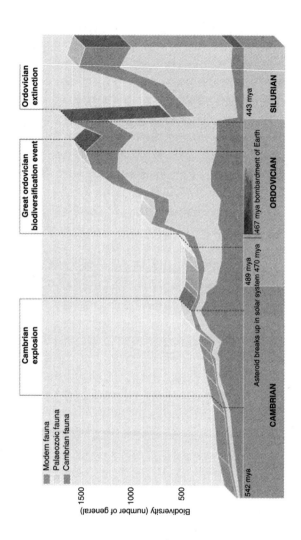

FIGURE 4.7 Life's big bangs: the Cambrian 'explosion' looks positively weedy in comparison with what happened in the Ordovician period.

# 5
## Nature's greatest inventions

*Evolution's methods are blind, brutish and aimless, yet it has fashioned some of the most exquisite machines in the known universe. And every now and then, it stumbles across a truly stunning innovation that rewrites the rules of life. Here are some of life's most incredible innovations.*

## *Multicellularity*

Ponder this one in the bath. Chances are you've just scrubbed your back with a choice example of one of evolution's greatest inventions. Or at least, a good plastic copy.

Sponges are a key example of multicellular life, an innovation that transformed living things from solitary cells into fantastically complex bodies. It was such a great move, it evolved at least 16 different times. Animals, land plants, fungi and algae all joined in.

Cells have been joining forces for billions of years. Even bacteria can do it, forming complex colonies with a three-dimensional structure and some division of labour. But hundreds of millions of years ago, eukaryotes – more complex cells that package up their DNA in a nucleus – took things to a new level. They formed permanent colonies in which certain cells dedicated themselves to different tasks, such as nutrition or excretion, and whose behaviour was well coordinated.

Eukaryotes could make this leap because they had already evolved many of the necessary attributes for other purposes. Many single-celled eukaryotes can specialize or 'differentiate' into cell types, dedicated to specific tasks such as mating with another cell. They sense their environment with chemical signalling systems, some of which are similar to those multicellular organisms use to coordinate their cells' behaviour. And they may detect and capture their prey with the same kind of sticky surface molecules that hold cells together in animals and other multicellular organisms.

So what started it? One idea is that clumping together helped cells avoid being eaten by making them too much of a mouthful for single-celled predators. Another is that single cells are often constrained in what they can do – for example, most cannot grow flagella to move and also divide at the same time. But a colony can both move and contain dividing cells if each cell in it takes its turn.

Researchers are now reconstructing the biology of the first multicellular creatures by studying the genomes of their nearest living relatives – for example, single-celled protozoans called choanoflagellates. This shines light on how animals came to evolve from them some 600 million years ago. Choanoflagellates and sponges – the only surviving witnesses to this step – share a common ancestor and choanoflagellates have a surprising number of equivalents to the signalling and cell-adhesion molecules unique to animals.

Yet bigger and more complex isn't necessarily better. Unicellular life still vastly outnumbers multicellular life in terms of both biomass and species numbers.

## *The eye*

Eyes appeared in an evolutionary blink and changed the rules of life forever. Before eyes, life was gentler and tamer, dominated by sluggish soft-bodied worms lolling around in the sea. The invention of the eye ushered in a more brutal and competitive world. Vision made it possible for animals to become active hunters, and sparked an evolutionary arms race that transformed the planet.

The first eyes appeared about 543 million years ago – the very beginning of the Cambrian period – in a group of trilobites called the Redlichia. Their eyes were compound, similar to those of modern insects, and probably evolved from light-sensitive pits. And their appearance in the fossil record is strikingly sudden – trilobite ancestors from 544 million years ago don't have eyes.

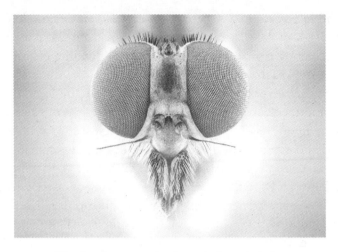

FIGURE 5.1 The development of eyes sparked an evolutionary arms race that transformed the planet.

So what happened in that magic million years? Surely eyes are just too complex to appear all of a sudden? Not so, according to Dan-Eric Nilsson of Lund University in Sweden. He has calculated that it would take only half a million years for a patch of light-sensitive cells to evolve into a compound eye.

That's not to say the difference was trivial. Patches of photosensitive cells were probably common long before the Cambrian, allowing early animals to detect light and sense what direction it was coming from. Such rudimentary sense organs are still used by jellyfish, flatworms and other obscure and primitive groups, and are clearly better than nothing. But they are not eyes. A true eye needs something extra – a lens that can focus light to form an image.

Trilobites weren't the only animals to stumble across this invention. Biologists believe that eyes could have evolved independently on many occasions, though genetic evidence suggests one ancestor for all eyes. But either way, trilobites were the first.

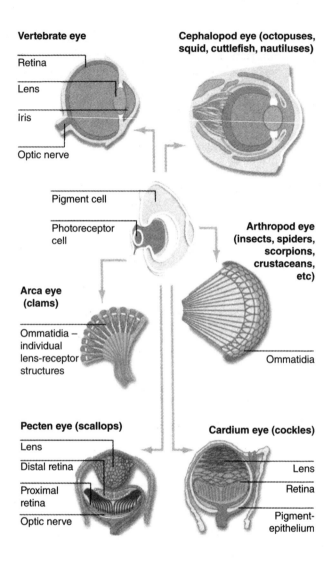

**Vertebrate eye**

Retina

Lens

Iris

Optic nerve

**Cephalopod eye (octopuses, squid, cuttlefish, nautiluses)**

Pigment cell

Photoreceptor cell

**Arthropod eye (insects, spiders, scorpions, crustaceans, etc)**

**Arca eye (clams)**

Ommatidia – individual lens-receptor structures

Ommatidia

**Pecten eye (scallops)**

Lens

Distal retina

Proximal retina

Optic nerve

**Cardium eye (cockles)**

Lens

Retina

Pigment-epithelium

FIGURE 5.2 Seeing the light: it was once believed that the eye had evolved independently on up to 65 occasions. But new genetic evidence suggests that it only happened once. The prototype eye (centre) then diversified into the myriad forms we see today.

And what a difference it made. In the sightless world of the early Cambrian, vision was tantamount to a super-power. Trilobites' eyes allowed them to become the first active predators, able to seek out and chase down food like no animal before them. And, unsurprisingly, their prey counter-evolved. Just a few million years later, eyes were commonplace and animals were more active, bristling with defensive armour. This burst of evolutionary innovation is what we now know as the Cambrian explosion.

However, sight is not universal. Of 37 phyla of multicellular animals, only six have evolved it, so it might not look like such a great invention after all – until you stop to think. The six phyla that have vision (including our own, chordates, plus arthropods and molluscs) are the most abundant, widespread and successful animals on the planet.

## The brain

Brains are often seen as a crowning achievement of evolution – bestowing the ultimate human traits such as language, intelligence and consciousness. But before all that, the evolution of brains did something just as striking: it lifted life beyond vegetation. Brains provided, for the first time, a way for organisms to deal with environmental change on a timescale shorter than generations.

A nervous system allows two extremely useful things to happen: movement and memory. If you're a plant and your food source disappears, that's just tough. But if you have a nervous system that can control muscles, then you can actually move around and seek out food, sex and shelter.

The simplest nervous systems are just ring-like circuits in cnidarians – the jellyfish, urchins and anemones. These might not be terribly smart, but they can still find the things they need and interact with the world in a far more sophisticated way than plants manage.

The next evolutionary step, which probably happened in flatworms in the Cambrian, was to add some sort of control system to give the movements more purpose. This sort of primitive brain is simply a bit of extra wiring that helps organize the networks.

Armed with this, finding food would have been the top priority for the earliest water-dwelling creatures. Organisms need to sort out nutritious from toxic food, and the brain helps them do that. Sure enough, look at any animal and you will find the brain is always near the mouth. In some of the most primitive invertebrates, the oesophagus actually passes right through the brain.

With brains come senses, to detect whether the world is good or bad, and a memory. Together, these let the animal monitor in real time whether things are getting better or worse. This in turn allows a simple system of prediction and reward. Even animals with really simple brains – insects, slugs or flatworms – can use their experiences to predict what might be the best thing to do or eat next, and have a system of reward that reinforces good choices.

The more complex functions of the human brain – social interaction, decision-making and empathy, for example – seem to have evolved from these basic systems controlling food intake. The sensations that control what we decide to eat became the intuitive decisions we call gut instincts. The most highly developed parts of the human frontal cortex that deal with decisions and social interactions are right next to the parts that control taste and smell and movements of the mouth, tongue and gut. There is a reason we kiss potential mates – it's the most primitive way we know to check something out.

## Language

As far as humans are concerned, language has got to be the ultimate evolutionary innovation. It is central to most of what makes us special, from consciousness, empathy and mental time travel to symbolism, spirituality and morality.

Exactly how our ancestors took this leap is one the hardest problem in science. Complex language – language with syntax and grammar, which builds up meaning through a hierarchical arrangement of subordinate clauses – evolved just once. Only human brains are able to produce language.

But why don't our close evolutionary relatives, chimps and other primates, have similar abilities? The answer could lie in neural networks unique to humans that allow us to perform the complex hierarchical processing required for grammatical language. These networks are shaped both by our genes and by experience. The first gene associated with language, FOXP2, was identified in 2001.

While humans and chimps both have the FOXP2 gene, the human and chimp versions are different and have different effects on the genes that FOXP2 targets in the brain. What's more, the brains of newborn humans are far less developed than those of newborn chimps, which means that our neural networks are shaped over many years of development immersed in a linguistic environment.

In a sense, language is the last word in biological evolution. That's because this particular evolutionary innovation allows those who possess it to move beyond the realms of the purely biological. With language, our ancestors were able to create their own environment – we now call it culture – and adapt to it without the need for genetic changes.

## Photosynthesis

Few innovations have had such profound consequences for life as the ability to capture energy from sunlight. Photosynthesis has literally altered the planet's face, transforming the atmosphere and cocooning Earth in a protective shield against lethal radiation.

Without photosynthesis, there would be little oxygen in the atmosphere, and no plants or animals – just microbes scratching a meagre existence from a primordial soup of minerals and carbon dioxide. It freed life from these constraints and the oxygen it generated set the stage for the emergence of complex life.

Before photosynthesis, life consisted of single-celled microbes whose sources of energy were chemicals such as sulphur, iron and methane. Then, around 3.4 billion years ago, or perhaps earlier, a group of microbes developed the ability to capture energy from sunlight to help make the carbohydrates they needed for growth and fuel. It is unclear how they achieved this feat, but genetic studies suggest that the light-harvesting apparatus evolved from a protein with the job of transferring energy between molecules. Photosynthesis had arrived.

But this early version of the process didn't make oxygen. It used hydrogen sulphide and carbon dioxide as its starting ingredients, generating carbohydrates and sulphur as end products. Some time later – just when is uncertain – a new type of photosynthesis evolved that used a different resource, water, generating oxygen as a by-product.

In those early days, oxygen was poisonous to life. But it built up in the atmosphere until some microbes evolved mechanisms to tolerate it, and eventually hit on ways to use it as an energy source. That was a pretty important discovery too: using oxygen to burn carbohydrates for energy is 18 times as efficient as doing it without oxygen.

Life on Earth became high-powered at this point, setting the scene for the development of complex, multicellular life forms – including plants, which 'borrowed' their photosynthetic apparatus from photosynthetic bacteria called cyanobacteria. Today, directly or indirectly, photosynthesis produces virtually all of the energy used by life on Earth.

As well as providing an efficient means to burn fuel, oxygen made by photosynthesis helps protect life. Earth is under constant bombardment from lethal UV radiation streaming out from the sun. A by-product of our oxygenated atmosphere is a layer of ozone extending 20 to 60 kilometres above Earth's surface, which filters out most of the harmful UV. This protective umbrella allowed life to escape from the sanctuary of the ocean and colonize dry land.

Now, virtually every biochemical process on the planet is ultimately dependent on an input of solar energy.

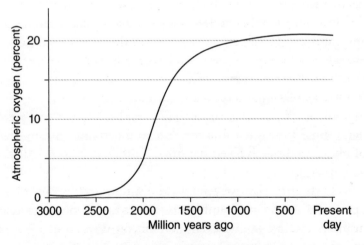

FIGURE 5.3 A world fit for animals: as photosynthesis took off it generated the oxygen we breathe.

## *Sex*

Birds do it, bees do it – for the vast majority of species, sexual reproduction is the only option. And it is responsible for some of the most impressive biological spectacles on the planet, from mass spawnings of coral so vast that they are visible from space, to elaborate sexual displays such as the dance of the bower bird, the antlers of a stag and – according to some biologists – poetry, music and art. Sex may even be responsible for keeping life itself going: species that give it up almost always go extinct within a few hundred generations.

Important as sex is, however, biologists are still arguing over how it evolved – and why it hasn't un-evolved. That's because, on the face of it, sex looks like a losing strategy.

Evolution ought to favour asexual reproduction for two reasons. First, in the battle for resources, asexual species should be able to outcompete sexual ones hands down. And secondly, because sperm and eggs contain only half of each parent's set of genes, an organism that uses sexual reproduction only gets 50 per cent of its genes into the next generation. Asexuals are guaranteed to pass on 100 per cent.

Clearly, though, there is something wrong with this line of reasoning. It's true that many species, including insects, lizards and plants, do fine without sex, at least for a while. But they are vastly outnumbered by sexual ones.

The enduring success of sex is usually put down to the fact that it shuffles the genetic pack, introducing variation and allowing harmful mutations to be purged (mutations are what eventually snuffs out most asexual species). Variation is important because it allows life to respond to changing environments, including interactions with predators, prey and – particularly – parasites. Reproducing asexually is sometimes compared to buying 100 tickets in a raffle, all with the same number. Far better to have only 50 tickets, each with a different number.

However useful sex may be now that we've got it, that doesn't tell us anything about how it got started. It could have been something as mundane as DNA repair. Single-celled, asexual organisms may have developed the habit of periodically doubling up their genetic material, then halving it again. This would have allowed them to repair any DNA damage by switching in the spare set. A similar exchange of DNA still happens during the production of eggs and sperm.

Parasites are also in the frame. Parasitic lengths of DNA known as transposons reproduce by inserting copies of themselves into the cell's normal genetic material. Imagine a transposon within a single-celled organism acquiring a mutation that happens to cause its host cell to periodically fuse with other cells before dividing again. The transposon for this primitive form of sex would be able to spread horizontally between many different cells. Once it arose in a population, parasitic sex would catch on pretty quickly.

## Death

Could evolution have brought the Grim Reaper into being? Yes, indeed. Not in all his guises, of course – living things have always died because of mishaps such as starvation or injury. But there's another sort of death in which cells – and perhaps, controversially, even whole organisms – choose annihilation because of the benefits it brings to some greater whole. In other words, death is an evolutionary strategy.

This is most obvious in the many varieties of programmed cell death or apoptosis, a self-destruct mechanism found in every multicellular organism. Your hand has five fingers because the cells that used to live between them died when you were an embryo. Embryos as tiny as 8 to 16 cells – just 3 or 4 cell divisions after the fertilized egg – depend on cell death: block apoptosis and development goes awry. Were it not for death, we would not even be born.

Even as adults we could not live without death. Without apoptosis we would all be overrun by cancer. Your cells are constantly racking up mutations that threaten to make your tightly controlled cell division run amok. But surveillance systems – such as the one involving the p53 protein, called the 'guardian of the genome' – detect almost all such errors and direct the affected cells to commit suicide.

Programmed cell death plays a central role in everyday life too. It ensures a constant turnover of cells in the gut lining and generates our skin's protective outer layer of dead cells. When the immune system has finished wiping out an infection, the now-redundant white blood cells commit suicide in an orderly fashion to allow the inflammation to wind down. And plants use cell death as part of a scorched-earth defence against pathogens, walling off the infected area and then killing off all the cells within.

It is easy to see how an organism can benefit from sacrificing a few cells. But evolution may also have had a hand in shaping the death of whole organisms. The cells of all higher organisms begin to age, or senesce, after just a few dozen cell divisions, ultimately leading to the death of the organism itself. In part that is one more protection against uncontrolled growth. But one controversial theory suggests this is part of an inbuilt genetic ageing program that sets an upper limit on all our lifespans.

Most evolutionary biologists reject the idea of an innate 'death program'. After all, they point out, animals die of old age in many different ways, not by one single route as apoptotic cells do. Instead, they view senescence as a sort of evolutionary junkyard: natural selection has little reason to get rid of flaws that appear late in life, since few individuals are lucky enough to make it to old age. But now that people routinely survive well past reproductive age, we suffer the invention evolution never meant us to find: death by old age.

## Parasitism

The name is synonymous with stealing, cheating and stealthy evil. But the age-old battle between parasites and their hosts is one of the most powerful driving forces in evolution. Without its plunderers and freeloaders, life would simply not be the same.

From viruses to tapeworms, barnacles to birds, parasites are among the most successful organisms on the planet, taking merciless advantage of every known creature. Take the tapeworm. This streamlined parasite is little more than gonads and a head full of hooks, having dispensed with a gut in favour of bathing in the nutrient-rich depths of its host's digestive system. In its average 18-year lifespan, a human tapeworm can generate 10 billion eggs.

Many parasites, such as the small liver fluke, have also mastered the art of manipulating their host's behaviour. Ants whose brains are infected with a juvenile fluke feel compelled to climb to the tops of grass blades, where they are more likely to be eaten by the fluke's ultimate host, a sheep.

The parasites that have had arguably the biggest effect on evolution are the smallest. Bacteria, protozoans and viruses can shape the evolution of their hosts because only the hardiest will survive infection. And humans are no exception: the genes for several inherited conditions protect against infectious disease when inherited in a single dose. For example, one copy of the gene for sickle cell anaemia protects against malaria. And it is still happening today. HIV and TB, for instance, are driving evolutionary change in parts of our genome, such as the immune-system genes.

Hosts can influence the evolution of their parasites too. For example, diseases which require human-to-human contact for transmission often evolve to be less deadly, ensuring a person will at least live long enough to pass it on.

Parasites can also drive evolution at a more basic level. Parasitic lengths of DNA called transposons, which can cut and paste themselves all over the genome, can be transformed into new genes or encourage the mutation and shuffling of DNA that fuels genetic variation. They have even been implicated in the origins of sex, as they may have driven selection for cell fusion and gamete formation.

---

### Other greats

Of course, there are many other evolutionary innovations that truly transformed life on earth. Nitrogen fixation is one of them. There would probably be no land plants and few land animals were it not for bacteria capable of converting inert, atmospheric nitrogen gas into organic compounds, and thereby making this essential element available to other life forms.

Another of evolution's great inventions is the one that enabled the development of the eye and the brain. Without the system of genes that defines front, back, top, bottom, left and right in organisms and oversees the folding of tissues into organs, life on Earth would surely be more like a form of slime.

---

## Superorganisms

A superorganism is a large number of individuals living together in harmony, achieving a better life by dividing their workload and sharing the fruits of their labours. We call this blissful state utopia, and have been striving to achieve it for at least as long as recorded history. Alas, our efforts so far have been in vain. Evolution, however, has made a rather better job of it.

Take the Portuguese man-of-war. It may look like just another jellyfish blob floating on the high seas, but zoom in with a microscope and you see that what seemed like one tentacled individual is in fact a colony of single-celled organisms. These 'siphanophores' have got division of labour down to a fine art. Some are specialized for locomotion, some for feeding, some for distributing nutrients.

This communal existence brings major advantages. It allows the constituent organisms, which would otherwise be rooted to the sea floor, to swim free. And together they are better able to defend themselves against predators, cope with environmental stress, and colonize new territory. Portuguese man-of-war jellyfish are true superorganisms.

With benefits like these on offer, it should come as no surprise that colonial living has evolved many times. Except that it does come with one big drawback, as the case of the slime bacteria, or myxobacteria, illustrates. These microbes are perhaps the simplest colonial organisms. Under normal circumstances individual bacteria glide along on lonely slime trails. Only when certain amino acids are lacking in their environment do individuals start to aggregate. The resulting superorganism consists of a stalk topped by a fruiting body containing spores. But since only the bacteria forming the spores will get the chance of dispersal and a new life, why should the others play along?

How this kind of cooperation evolved, and how cheats are prevented from taking advantage of it remains unclear for some types of colonial life.

But in one group of animals, the colonial insects, we do know what the trick is – and it's an ingenious one. Females develop from fertilized eggs, while males develop from unfertilized ones. This way of determining sex, called haplodiploidy, ensures that sisters are more closely related to each other than to their own offspring. And this means that the best chance they can give their own genes of surviving is to look after each other rather than lay eggs of their own. This is what provides

FIGURE 5.4 The Portuguese man-of-war looks like a jellyfish but is actually a commune for single-celled organisms.

the stability at the heart of the beehive and termite mound, and in many other insect colonies where haplodiploidy has evolved at least a dozen times.

True sociality, or **eusociality** as it is technically known, is found in all ants and termites, in the most highly organized bees and wasps, and in some other species, not all of which employ haplodiploidy. And although these mini societies need careful policing to keep cheats at bay, this is probably the closest thing on Earth to utopia.

---

### Evolution's mistakes

Evolution can fall well short of perfection too. Here are some examples:

- **The female pelvis:** Human adaptation to walking upright has made giving birth more dangerous for women than for any other primate.
- **Linear chromosomes:** The ends of linear chromosomes erode as cells divide, something that cannot happen with circular chromosomes.
- **Mutant glo gene:** Like most primates, humans cannot make vitamin C thanks to mutations in the L-gulono-Y-lactone oxidase (GLO) gene, rendering us vulnerable to scurvy unless we get plenty in our diet.
- **Windpipe:** Positioned next to the gullet, it means choking is not uncommon.
- **Vulnerable brain cells:** A few minutes of oxygen deprivation causes permanent brain damage in humans, yet an epaulette shark can survive for over an hour without oxygen
- **Odontoid process:** This extension of the last neck vertebra can easily fracture and damage the brainstem.

---

> • **Feet:** After coming down from the trees, we ended up walking on the 'wrists' of our lower limbs, leading to all sorts of structural weaknesses.
> • **The Y chromosome:** It is gathering mutations because it can't swap DNA with the X chromosome.

## Symbiosis

Crocodiles with gleaming gums, coral reefs, orchids, fish with glow-in-the-dark lures, ants that farm. All these result from swapping food – for cleaning services, for transport, for sunscreen, for shelter, and of course for other food.

Symbiosis has many definitions, but we'll take it to mean two species engaging in physically intimate, mutually beneficial dependency, almost invariably involving food. Symbiosis has triggered seismic shifts in evolution, and evolution in turn continually spawns new symbiotic relationships.

Perhaps the most pivotal couplings were the ones that turbo-charged complex, or eukaryotic, cells. Eukaryotes use specialized organelles such as mitochondria and chloroplasts to extract energy from food or sunlight. These organelles were originally simpler, prokaryotic cells that the eukaryotes engulfed in an eternal symbiotic embrace. Without them life's key developments, such as increasing complexity and multicellular plants and animals, would not have happened.

Symbiosis has popped up so frequently during evolution that it is safe to say it's the rule, not the exception. Angler fish in the deep ocean host bioluminescent bacteria in appendages that dangle over their mouths. Smaller fish lured by the light are easy prey. At the ocean surface, coral polyps provide homes for photosynthetic algae, and swap inorganic waste products for organic carbon compounds – one reason why nutrient-poor tropical waters can support so much life. The algae also produce a chemical that absorbs ultraviolet light and protects the coral.

More than 90 per cent of plant species are thought to engage in symbiotic couplings. Orchid seeds, for example, are little more than dust, containing next to no nutrients. To germinate and grow, they digest a fungus that infects the seed.

Plovers pick leeches from crocodiles' teeth, offering dental hygiene in return for food. Leafcutter ants use chopped-up leaves as a fertilizer for the fungus they grow in underground chambers. The ants cannot digest the leaves but the fungus that feeds on them produces a tasty meal of sugars and starch while breaking down the toxins in the leaves. And there is not an animal out there, including us, that can survive without the bacteria that live in its gut, digesting food and producing vitamins.

## But ... nature is not endlessly creative

It often seems that nature invented pretty much everything that can be invented long before humans arrived on the scene – including the wheel, kind of (see 'Wheels are a pretty effective method of getting around, so why did they never evolve in nature?'). Nevertheless, there are structures that would clearly be useful but have never evolved – and could probably never evolve, at least not on Earth.

Zebras with built-in machine guns would rarely be bothered by lions, some point out. So why can evolution invent some things but not others?

This is an extremely difficult issue to tackle: how can we study something that has not happened? One way to approach it is to start with a question used by those who deny evolution and believe that many of nature's inventions, such as the eye or the bacterial flagellum, are simply too complex to have evolved. What use is half a wing, they ask?

Very useful, it turns out. The wings of insects might have evolved from flapping gills that were originally used for rowing on the surface of water. This is an example of exaptation – structures and behaviours that evolved for one purpose but take on a wholly new one, while remaining useful at every intermediate stage.

Turn this argument around, however, and it suggests that some features cannot evolve because a half-way stage really would be of no use. For example, two-way radio might be useful for many different animals, for making silent alarm calls or locating other members of your species. So why hasn't it evolved? The recent invention of nanoscale radio receivers suggests it is not physically impossible.

The answer might be that half a radio really is useless. Detecting natural radio waves – from lightning, for instance – would not tell animals anything useful about their environment. That means there will be no selection for mutations that allow organisms to detect radio waves. Conversely, without any means of detecting radio waves, emitting them would serve no useful purpose. Radar might not be able to evolve for similar reasons.

The contrast with visible light could hardly be greater. It is clear that simply detecting the presence or absence of light would be advantageous in many environments, that even a blurry picture is better than nothing at all, and so on right up to hawk-eyed sharpness.

## Seaweed skies

Emitting visible light can be helpful too, even for creatures that cannot detect it themselves. For the bioluminescent phytoplankton that light up ocean waves, for instance, it is a way of summoning predators that eat the phytoplankton's enemies. A similar argument applies to sound: it is not hard to see how forms of echolocation evolved independently in groups such as bats, cave swiftlets and whales.

One might also wonder why plants that float in the sky like balloons have never evolved. The idea does not seem too far-fetched at first glance: many seaweeds have floats called pneumatocysts, filled with oxygen or carbon dioxide. Other algae can produce hydrogen. So fill a large, thin pneumatocyst with hydrogen and perhaps a seaweed could fly. Flying plants would beat water and land plants to the light, giving them a big advantage, so why aren't our skies filled with living green balloons?

Perhaps partly because large pneumatocysts with extremely thin membranes would be far more vulnerable to predators and damage from waves, so an intermediate stage could never evolve. What's more, algae produce hydrogen only when there's a lack of sulphur in the water, and in any case the molecules of hydrogen gas are so tiny that they would leak out of any pneumatocyst. Half a hydrogen balloon doesn't look very good for anything, at least on our planet. Even evolution has its limits.

---

**Wheels are a pretty effective method of getting around, so why did they never evolve in nature?**

It's not true to say that nature hasn't invented the wheel: bacteria have been using it to get around for millions of years. It is the basis for the bacterial flagellum, which looks a bit like a corkscrew and which rotates continuously to drive the organism along. About half of all bacteria have at least one flagellum. Each is attached to a 'wheel' embedded in the cell membrane that rotates hundreds of times per second, driven by a tiny electric motor. It is a sophisticated piece of nanotechnology and even has a reverse gear.

So, far from nature not having invented the wheel, given the very large number of bacteria in existence, there are probably more wheels in the world than any other form of locomotion.

There are also macroscopic life forms that roll like a wheel. Tumbleweed, for example. Then there is a salamander living in the Californian mountains that coils itself up and rolls downhill when threatened. The pearl moth caterpillar goes one better and can roll itself along a flat surface for four or five revolutions to escape predators.

# 6
# Myths and misconceptions

*For those who have never had the opportunity to learn much about biology or science in general, claims about evolutionary theory made by those who believe in supernatural alternatives can appear convincing. Even among those who do accept the reality of evolution, misconceptions still abound. Most of us are happy to admit that we do not understand, say, quantum mechanics in physics, yet we would baulk at saying the same about evolution. In fact, as biologists are discovering, evolution can be stranger than their predecessors ever imagined.*

# Evolution: a guide for the perplexed

Here is a primer to a few common myths and misconceptions about evolution.

## Everything is adaptation

*Contrary to popular belief, not all characteristics of plants and animals are adaptations or the result of natural selection.*

Why do so many of us spend our evenings in front of the TV with a microwave meal? Could it be that television is the modern equivalent of a Neolithic fire, making TV dinners 'the natural consequence of hundreds of thousands of years of human evolution', as a researcher once concluded?

Stop laughing. It is very tempting to invent evolutionary 'just so' stories to explain almost any aspect of our body or behaviour. We all tend to assume that everything has a purpose – but we are often wrong.

Take male nipples. Male mammals clearly do not need them. They have them because females do: it doesn't cost much to grow a nipple, so there has been no pressure for the sexes to evolve separate developmental pathways, to switch off nipple growth in males. Some researchers claim the female orgasm exists for the same reason, though this is far more controversial.

Or consider your sense of smell. Do you find the scent of roses overwhelming or struggle to smell anything at all? Can you detect the distinctive odour that most people's urine acquires after eating asparagus? People vary greatly when it comes to smell, and this is probably less to do with natural selection than with chance mutations in the genes coding for the smell receptors.

Then there are features which do result from selection, but for another trait entirely. For instance, the short stature of pygmies might have no survival advantage in itself, but instead be a side effect of selection for early childbearing in populations where mortality is high. Similarly, since the same gene often has different roles at different times of development or in different parts of the body, selection for a variant that is beneficial in one way can have other, seemingly unrelated effects. Male homosexuality might be a side effect of genetic variants that boost female fertility. What's more, a mediocre or even poor gene variant can spread rapidly through a population if it happens to be located near a highly beneficial gene.

Other features of plants and animals, such as the wings of ostriches, are adaptations no longer needed for their original purpose. These vestigial traits can persist because they make no difference to an individual's chances of survival, or they have taken on another function, or because even though they have become disadvantageous, they occur in a population that is too small or has undergone too few generations for evolution to eliminate them.

A prime example in humans is the appendix. While claims abound that it has this or that function, the evidence is clear: you are more likely to survive without an appendix than with one. Another example is wisdom teeth. Having a smaller, weaker jaw allowed our ancestors to grow larger brains, but left less room for molars. Yet many of us still grow teeth for which there is no room, and the consequences can be fatal.

Evolutionary psychology in particular is notorious for attempting to explain every aspect of human behaviour, from gardening to rape, as an adaptation that arose when our ancestors lived on the African savannah. Some behaviours may indeed be past adaptations, but in the absence of any proof, claims about TV dinners should be taken with a large pinch of salt.

## Evolution can't be disproved

*There are all sorts of findings and experiments that could have falsifed evolution, but in the century and a half since Darwin published his theory, not a single one has done so.*

When asked what would disprove evolution, the biologist J.B.S. Haldane famously growled: 'Fossil rabbits in the Precambrian.' What he meant was that evolution predicts a progressive change over time in the millions of fossils unearthed around the world: multicellular organisms should come after unicellular ones; jawed fish should come after jawless ones, and so on. All it would take is one or two exceptions to challenge the theory. If the first fossil amphibians were older than the first fossil fish, for example, it would show that amphibians could not have evolved from fish. No such exceptions have ever been found anywhere.

The discovery of a mammal–bird hybrid, such as a feathered rabbit, could also disprove evolution. There are animals with a mixture of mammalian and reptilian features – such as the spiny anteater – and there are fossils with a mixture of bird and reptilian features, such as the toothy archaeopteryx. But no animals have a mixture of mammalian and bird features. This is exactly what you would expect if birds and mammals evolved from separate groups of reptiles, whereas there is no reason why a 'designer' would not have mixed up these features, creating mammals with feathers and bird-like lungs, or furry, breast-feeding ostriches.

A young Earth would also be a problem for evolution, since evolution by natural selection requires vast stretches of time – 'deep time' – as Darwin realized. Some thought evolution had been falsified in the nineteenth century when physicist William Thomson calculated that the Earth was just 30 million years old. In fact, several lines of evidence, such as lead isotopes, show the Earth is far older than even Darwin imagined – about 4.5 billion years old.

Suppose for a moment that life was designed rather than having evolved. In that case, organisms that appear similar might have very different internal workings, just as an LCD screen has a quite different mechanism to a plasma screen. The explosion of genomic research, however, has revealed that all living creatures work in essentially the same way: they store and translate information using the same genetic code, with only a few minor variations in the most primitive organisms. Huge chunks of this information are identical or differ only slightly even between species that appear very different.

What's more, the genomes of complex creatures reveal a lack of any intelligence or foresight. Your DNA consists largely of millions of defunct copies of parasitic DNA. The inescapable conclusion is that if life was designed, the designer was lazy, stupid and cruel.

Not only that, if organisms had been designed for particular roles, they might be unable to adapt to changing conditions. Instead, countless experiments, both planned and unplanned, show that organisms of all kinds evolve when their environment is altered, provided the changes are not too abrupt. In the laboratory, tweaking organisms' environments has enabled researchers to produce bacteria, plants and animals with all kinds of novel characteristics – even entirely new species. In the wild, human activity is reshaping many species: urban birds are diverging from their country cousins, some fish are getting smaller because fishermen keep only big fish, and trophy hunting is turning bighorn sheep into smallhorns, for instance.

## Natural selection leads to ever greater complexity

*Actually, natural selection can lead to ever greater simplicity, and complexity may initially arise when selection is weak or absent.*

Use it or lose it. That old adage applies to evolution as well as everyday life, and explains why cave fish are eyeless and parasitic tapeworms gutless.

Until recently, such examples were considered the exception, but it seems we may have seriously underestimated the extent to which evolution likes to simplify matters. There are entire groups of apparently primitive creatures that are turning out to be the descendants of more complex organisms. For instance, the ancestor of brainless starfish and sea urchins had a brain; why their descendants dispensed with a brain is still unclear.

Despite this, there is no doubt that evolution has produced ever more complex life forms over the past 4 billion years. This is usually assumed to be the result of natural selection, but recently some biologists studying our bizarre and bloated genomes have turned this idea on its head. They propose that, initially at least, complexity arises when selection pressure is weak or absent. How could this be?

Suppose an animal has a gene with two different functions. As a result of mutation some offspring may get two copies of this gene. In a large population where competition is fierce and selection pressure strong, such mutations are likely to be eliminated because they do not increase an individual's fitness and are probably slightly disadvantageous.

In smaller populations where selection pressure is weak, however, these mutations have a small chance of surviving and spreading as a result of random genetic drift. If this happens, the duplicate genes will start to acquire mutations of their own. A mutation in one copy might destroy its ability to carry out the first of the original gene's two functions, while the other copy might lose the ability to perform the second function. Again these changes don't confer any advantage – such animals would still look and behave exactly the same – but these mutations might also spread by genetic drift. So the population would have gone from having one gene with two functions to two genes with one function each.

This increase in genomic complexity would have occurred not because of selection pressure but despite it. Yet it can be the foundation of greater physical or behavioural complexity because each gene can now evolve independently. For example, either can be switched on or off at different times or in different tissues. And as soon as any beneficial mutations arise, natural selection will kick in.

It seems there are opposing pressures at the heart of evolution: while complex structures and behaviours, such as eyes and language, are undoubtedly the product of natural selection, strong selection – as in large populations – blocks the random genomic changes that can throw up greater complexity in the first place.

## Evolution produces perfection

*You don't have to be perfectly adapted to survive, you just have to be as well adapted as your competitors are.*

It's a theme endlessly repeated in wildlife documentaries. Again and again we are told how perfectly animals are adapted to their environment. It is, however, seldom true.

Take the red squirrel, which appeared to be perfectly adapted to its environment until the grey squirrel turned up in the UK and demonstrated that it is in fact rather better adapted to broadleaf forests.

There are many reasons why evolution does not produce perfect 'designs'. Natural selection only requires something to work, not to work as well as it could. Botched jobs are common. The classic example is the panda's 'thumb', a modified wrist bone that the animal uses like an opposable thumb to grasp bamboo. It's far from the ideal tool for the job, but since

FIGURE 6.1 Evolution has crafted many structures from the fins of early fish, including wings, hoofs and hands.

the panda's true thumb is fused into its paw, the panda had to settle for a clumsier alternative.

Evolution is far more likely to reshape existing structures than throw up novel ones. The lobed fins of early fish have turned into structures as diverse as wings, hoofs and hands. What this means is that we have five fingers because amphibians had five digits, not because five fingers is necessarily the optimal number for the human hand.

Many groups haven't evolved features that would make them better adapted. Sharks lack the gas bladder that allows bony fish to precisely control their buoyancy, and instead have to rely on swimming, buoyant fatty livers and, occasionally, gulping air. Mammals' two-way lungs are far less efficient than those of birds, in which the air flows in one direction.

Continual mutation also means that potentially useful features can get lost. Many primates cannot make vitamin C, an ability that wasn't missed in animals that get lots of vitamin C in their diet. However, such losses can be limiting if the environment changes, as one primate discovered on long sea voyages.

Evolution's lack of foresight also leads to inherently flawed designs. The vertebrate eye, with its blind spot where the wiring goes through the retina, is one example. Once natural selection fixes upon a bad – but workable – design, a species' descendants are usually stuck with it.

Environments also change. In the arms race between predator and prey, parasite and host, species have to keep evolving just to maintain their current level of fitness, let alone get even fitter. As the Red Queen says in *Through the Looking Glass*: 'It takes all the running you can do, to keep in the same place.'

Humans aren't running fast enough. Evolving and adapting is a numbers game: the larger a population and the more generations there are, the more mutations will appear and the more chances there will be for natural selection to favour the beneficial and eliminate the harmful. Around 10 billion new viral particles can be produced every day in the body of a person infected with HIV; the total human population on Earth was no more than a few million until fairly recently.

A bacterium can produce 100,000 generations in a decade, but there have probably been fewer than 25,000 generations since the human lineage split from that of chimpanzees. So it's hardly surprising that in less than a human lifespan, we've seen the evolution of new viruses, such as HIV.

Our evolution has accelerated in the last 10,000 years, but we are changing our environment even faster, leading to problems ranging from obesity and allergies to addictions and shortsightedness. Viruses and bacteria might approach perfection: we humans are at best a very rough first draft.

# Evolutionary science is not predictive

*We cannot say exactly what life will look like in a billion years, but evolutionary theory can make a few predictions.*

Cosmologists make precise predictions about what will happen to the universe in 20 billion years' time. Biologists struggle to predict how a few bacteria in a dish might evolve over 20 hours. Some claim that this lack of precise predictive power means evolution is not scientific.

However, what matters in science is not how much you can predict on the basis of a theory or how precise those predictions are, but whether you can make predictions that turn out to be right. Meteorologists don't reject chaos theory because it tells them it is impossible to predict the weather 100 per cent accurately – on the contrary, they accept it because weather follows the broad patterns predicted by chaos theory.

The difficulty in predicting the path of evolution partly springs from organisms' freedom to evolve in quite different directions. If we could wind the clock back 4 billion years and let life evolve all over again, its course might well be different. Life on this planet has also been shaped by chance events. If an asteroid had not wiped out the dinosaurs, intelligent life might have been very different, if it evolved at all.

Nevertheless, although evolution's predictive power might appear limited, the theory can and is used to make all sorts of predictions. For a start, Darwin predicted that transitional fossils would be discovered, and millions – trillions if you count microfossils – have been uncovered. What's more, researchers have predicted in which kinds of rocks and from what eras certain transitional fossils should turn up in, then gone out and found them, as with the half-fish, half-amphibian Tiktaalik.

Or take the famous peppered moth, which evolved black colouring to adapt to pollution-stained trees when industrialization took place. Remove the pollution and, evolutionary theory predicts, the light strain should once again predominate – which is just what is happening.

This predictive power can also be put to much more practical use. For instance, evolutionary theory predicts that if you genetically engineer crops to produce a pesticide, this will lead to the evolution of insect strains which resist that pesticide, but it also predicts that you can slow the spread of resistance genes by growing regular plants alongside the GM ones. That has proved to be the case. Many researchers developing treatments for infectious diseases try to predict how resistance might evolve and to find ways to prevent this from happening, such as prescribing certain drugs in combination. This slows the evolution of resistance because pathogens have to acquire several different mutations to survive the treatment.

## Natural selection is the only means of evolution

*Much change is due to random genetic drift rather than positive selection. It could be called the survival of the luckiest.*

Take a look in the mirror. The face you see is rather different from that of a Neanderthal. Why? The answer could be genetic drift. With features such as the shape of your skull, which can vary in form with little change in function, chance might play a bigger role in evolution than natural selection.

DNA is under constant attack from chemicals and radiation, and errors are made when it is copied. As a result, each human embryo contains 100 or more new mutations. Natural selection will eliminate the most harmful – those that kill the embryo, for instance. Most mutations make no difference because they occur in junk DNA, which makes up the vast majority of our genome. A few cause minor changes that are neither particularly harmful nor beneficial.

While most new neutral mutations die out, a few spread through later generations purely by chance. The odds of this happening are tiny, but the sheer number of mutations that arise make genetic drift a significant force. The smaller a population, the more powerful it is (see Figure 6.2).

Population bottlenecks have the same effect. Imagine an island where most mice are plain but a few have stripes. If a volcanic eruption wipes out all the plain mice, striped mice will repopulate the island. It's survival of the luckiest, not the fittest.

These processes have almost certainly played a big role in human evolution. Human populations were tiny until around 10,000 years ago, and genetic evidence suggests that we went through a major bottleneck around 2 million years ago.

Most of the genetic differences between humans and other apes – and between different human populations – are due to genetic drift rather than selection, but as most of these mutations are in the nine-tenths of our genome that is junk, they do not make any difference. Of those that do affect our bodies or behaviour, it is likely that at least a few have spread because of drift rather than selection.

## Half a wing is no use

*Just as objects designed for one purpose can be used for another, so genes, structures and behaviours that evolve for one purpose become adapted to do another.*

What use is half a wing? It's a question that those who doubt evolution first asked more than a century ago. When it comes to insects, rowing and skimming could be the answer. Stonefly nymphs have flapping gills for extracting oxygen from water. When standing on the water's surface, early insects could have used these gills for getting oxygen and propulsion rowing

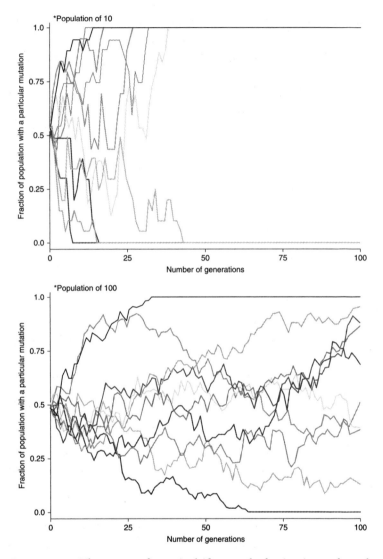

FIGURE 6.2 The power of genetic drift: natural selection is not the only force in evolution. Mutations that have little or no effect on fitness can spread throughout a population or die out due to chance alone. Each graph shows ten simulation runs from the same starting point.

simultaneously. Some stoneflies still stand on the surface and 'row' across water using their wings.

Over time, flapping could have replaced rowing as the main means of propulsion, allowing insects to skim across the water's surface: low levels of friction on this scale mean proto-wings would not have had to generate much air flow to be useful for skimming.

As these proto-wings became more efficient and specialized, early insects may have taken further steps towards flying. While some skimming insects keep all six legs on the water's surface, faster skimmers keep just four legs or two legs on the water. This surface-skimming hypothesis concerning the evolution of insect flight shows how flapping gills could gradually have turned into wings while remaining useful at every stage.

What about the wings of birds? In some dinosaurs, the scales covering their bodies evolved into hair-like feathers, most likely to insulate warm-blooded bodies or help keep eggs warm. Those dinosaurs with feathers on their limbs might then have started to exploit the aerodynamic properties offered by feathers, perhaps gliding between trees or running faster along the ground. Fossils show a gradual transition from downy, hair-like feathers into the rigid flight feathers that form the key part of birds' wings.

Another idea that is gaining favour is that flapping forelimbs helped the ancestors of birds to run up steep slopes or climb trees – a technique many birds still employ today.

Without a time machine it is difficult to prove exactly what early birds or insects used 'half a wing' for. But it is now clear that half a wing can have all sorts of uses. Indeed, there are numerous examples of physical structures and behaviours that evolved for one purpose acquiring another one, a process called exaptation.

Evo-devo – evolutionary developmental biology – is even starting to identify the precise mutations that underlie such changes. For instance, the forelimbs of the ancestor of bats turned into wings partly thanks to a change in a gene called BMP2 that made its 'fingers' far longer than normal (see Chapter 7).

The webbing between the extra-long digits that makes up the bat wing is a reappearance of a long-lost feature: as embryos, all tetrapods initially develop webbed digits, a hangover from our fish ancestors. Normally, this webbing kills itself off at an early stage, but in bats this cell suicide is blocked.

Repurposing a structure does not have to involve the loss of the original structure. Reptilian jaw bones turned into mammalian ear bones, without the loss of the jaw. The neural circuitry that allows us to make fine limb movements may have been adapted to produce speech as well.

In fact, almost every feature of complex organisms can be seen as a variation on a theme. Switching off one gene in fruit flies, for instance, can turn their antennae into legs.

Sometimes just one aspect of a feature can be co-opted for another use. The first hard mineralized structures to evolve in our ancestors were the teeth of early fishes known as conodonts. Once the ability to form hard hydroxyapatite had evolved, it could be exploited elsewhere in the body and may have been the basis of the bony skeletons of all vertebrates.

There are all kinds of routes by which structures and behaviours that evolved for one purpose can contribute to new structures and abilities. Just because it is not immediately obvious how something as complex as a bacterial flagellum does not prove it did not evolve.

## But isn't evolution just a theory?

This is a common question posed by creationists.

And yes, evolution is a theory, just like Einstein's theory of special relativity. By theory, scientists mean an explanation backed by evidence. What creationists mean is that evolution is just a hypothesis, unsupported by evidence – which it is not. Sure, there are plenty of details to fill in. But would you jump off a skyscraper on the basis that the clash between general relativity and quantum theory means there are serious problems with our theory of gravity? It makes no more sense to question the reality of evolution because scientists are still debating about some of its finer aspects than it does to question the existence of gravity for the same reason. As surely as dropped objects fall, life has evolved and continues to do so.

# 7
## Digging deeper

*The basic principles of evolution by natural selection, as outlined by Darwin and Wallace more than 150 years ago, have proved incredibly robust. However, many details of the underlying mechanisms have only been recently uncovered, shining new light on the process of evolution. For a start, it doesn't always proceed at the glacially slow pace that Darwin envisaged. New techniques are revealing the underlying mechanisms for evolutionary innovation and how species are formed. Evolution can even be tested in the lab.*

## *Evolution in the fast lane*

The fossil record and genetic studies suggest that evolution happens at a glacially slow pace, but that couldn't be further from the truth.

Take, for example, the stickleback that Michael Bell found back in 1990, when he was driving past Loberg Lake in Alaska. Bell, a biologist who studies the evolution of sticklebacks, had not planned to collect any fish, as the native sticklebacks had been exterminated in 1982 to improve the lake for anglers.

To Bell's surprise, they found that marine sticklebacks had recolonized the lake. This in itself was not all that unusual: marine sticklebacks can live in fresh water, and most freshwater species are descended from marine ones that colonized streams and lakes as the ice retreated at the end of the last ice age.

But there was something odd about these sticklebacks. Ten thousand years on from the ice age, freshwater sticklebacks are quite different from their sea-going ancestors. The most obvious change is loss of armour plates, which seem to take longer to develop in fresh water. In lakes, lightly armoured fish may outgrow and outcompete fully armoured fish.

This trait was assumed to evolve slowly, over thousands of years, so Bell was surprised to find that some of the fish he caught in Loberg Lake had fewer plates. In 1991 he asked a friend to collect some more fish. Sure enough, more had lost their armour.

Bell, who is based at Stony Brook University in New York, began collecting sticklebacks every year. Each time, he found more lightly armoured fish. By 2007, 90 per cent were of the low-armour form. Far from taking millennia, the trait had evolved in a couple of decades (see Figure 7.1).

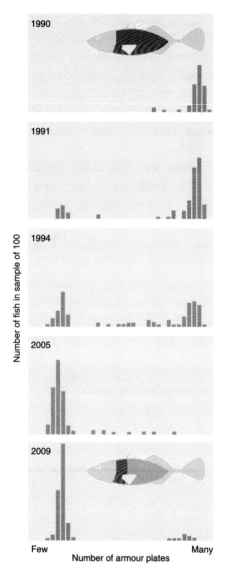

FIGURE 7.1 Ready, steady, evolve! Freshwater sticklebacks are known to evolve from heavily armoured marine forms. Samples from Loberg Lake in Alaska show that it can happen in less than 20 years.

Compared with the gradual process described by Darwin, this is evolution at warp speed. What is really startling, though, is that far from being exceptional, high-speed evolution is starting to look like the norm. Very few biologists set out to look for evidence of ongoing evolution, but wherever they do, they find it – from weeds and pests to fish to humans. It now appears that whenever the environment changes in any way, life evolves. Fast.

Such findings throw up a paradox. The two main ways to study evolution – the fossil record and comparisons of genomes of living organisms – suggest that the process is gradual, with some species barely changing over tens of millions of years. If evolution is as rapid as some biologists now claim, how come the fossil record and genetic studies suggest it is usually very slow?

## Rapid evolution

Reports of rapid evolution go back a surprisingly long way. It turns out that British entomologist Albert Farn wrote to Darwin in 1878 to point out that darkly coloured annulet moths were becoming more common than lighter moths in areas blackened by pollution. This was nearly 20 years before it was first suggested that the famous peppered moths were turning black for the same reasons.

In 1897, it emerged that several insect populations were becoming resistant to insecticides. By the 1930s, more examples had surfaced, such as scale insects developing resistance to hydrogen cyanide.

Over the following decades, biologists stumbled upon more and more examples. A few became famous, such as the peppered moth, but all were regarded as curiosities. 'People went, "wow, that's amazing, that must be the exception",' says Michael Kinnison of the University of Maine in Orono, one of the first researchers to set out specifically to look at evolution in action.

Today, there are probably thousands of examples, and a growing number of biologists think that far from being an exception, rapid evolution is common. And thanks to advances in genetics, we are beginning to understand how it is possible.

Bell's stickleback record remains one of the best-documented examples. Besides losing armour, the fish have also acquired other traits typical of freshwater fish, such as smaller gills. Their immune systems have also evolved to cope with different threats. Research from earlier this year even shows that a population of stickleback fish in Lake Constance, Switzerland, is splitting into two species before our eyes. The one living in the main lake has longer spines and tougher armour compared with the other, which lives in the streams flowing into the lake.

From genetic studies we know that armour loss is due to mutations in a gene called EDA, which plays a role in skin development. These mutations are also found in marine sticklebacks, although they are very rare. They persist at low levels because the trait is recessive, meaning fish lose their armour only if they inherit two mutant copies of the gene.

But once the sticklebacks move into fresh water where less armour is an advantage, the mutations are desirable and rapidly become more common as natural selection does its work. This explains how the same trait evolved repeatedly as sticklebacks colonized lakes after the ice age.

Such pre-existing genetic diversity seems to be what allows populations to evolve rapidly. Support for this idea comes from a study of sticklebacks in Cook Inlet, Alaska, which recently switched to living in fresh water. The armour of these fish has hardly changed at all, and Bell's team has found that they are less genetically diverse than those in Lake Loberg.

While rapid evolution usually involves existing mutations, new ones can play a role too. For instance, the mosquito *Culex pipiens* evolved resistance to organophosphate insecticides when an unusual mutation produced several copies of one gene, enabling it to make more of an enzyme that could break the pesticides down. This new mutation has spread worldwide.

## New species in next to no time

In the right circumstances, even new species can evolve in next to no time. In 1866, farmers in the US reported that an unknown maggot was attacking their apples, a crop introduced two centuries earlier. Entomologist Benjamin Walsh suggested that the 'apple maggot' was a strain of the native hawthorn fly that had switched diets. Walsh had previously suggested this kind of process could lead to speciation.

We now know that Walsh was right. Genetic studies have shown that the hawthorn fly appears to be in the process of splitting into two species. What's more, the parasitic wasps whose larvae feed on the maggots are also splitting into two species.

More examples keep turning up. A species of fish in a lake in Nicaragua has split in two in only 100 years. The new variety has evolved a narrower, pointier head and fatter lips, ideal for nibbling insects from crevices. The original variety has sturdier jaws and extra teeth to crack snail shells. Lab studies suggest the strains do not mate with each other even when put together, which would mean they are on their way to becoming separate species.

Yet another example comes from the famous Galapagos finches. Since 1973, husband-and-wife team Peter Grant and

Rosemary Grant have been studying the finches on the island of Daphne Major, in one of the few long-term studies of ongoing evolution. They reported in 2010 that a new species of finch might be evolving. In 1981, a medium ground finch (*Geospiza-fortis*) from another island reached Daphne Major and interbred with the local birds, producing offspring with unusual beaks and songs. After four generations, following a severe drought that killed many birds, this new strain stopped interbreeding with the other finches. It's not clear why interbreeding stopped, but if the birds continue to shun the locals they will become a new species.

As the list of examples grew, Kinnison and his colleagues began to pull them together and look at what they tell us about evolution. 'We started to realize that maybe this was not the exception, that this was the norm.' In fact, he now argues that the term 'rapid evolution' is misleading, because it implies evolution is normally slow. Instead, he and his colleagues prefer 'contemporary evolution'. Of course, proving that contemporary evolution is the norm in a world of millions of species is a challenge.

If rapid evolution really is the norm, how come fossil and genetic studies suggest it is slow? The answer may be that new species and traits not only evolve rapidly, they also disappear fast too and do not leave their mark on the fossil or genetic record.

## Evolution in reverse

The best example also comes from the Galapagos. In 1977 a drought on Daphne Major wiped out plants that produce small seeds, and many of the ground finches that fed on them died. Those with larger beaks that enabled them to feed on bigger seeds did better, and within a few generations beak size had increased by around 4 per cent. The wet year of 1983 saw small seeds become abundant again and soon beak size had shrunk again – evolution had gone into reverse.

Speciation can also go into reverse. On the nearby island of Santa Cruz, two incipient species are collapsing back into one. Work in the 1960s showed the finches on this island had split into large-and small-beaked strains, specializing in different-sized seeds. Now most have medium-sized beaks, probably as a result of people feeding birds rice, making small or large beaks less of an advantage.

Many other examples are being discovered. Lake Victoria in east Africa is home to more than 500 species of cichlid fish, many of which split off in the past 15,000 years. Now many species are merging back together. The reason is that females recognize males of the same species by their bright colours. As the lake has become murkier due to human activity, females are increasingly breeding with the wrong males, giving rise to hybrids that eventually replace the two original species.

This evolutionary toing and froing may well be the norm. As a result of fluctuating selection pressures, populations probably evolve rapidly in one direction and then the other, ending up back where they started.

Evolutionary yo-yoing can also be driven by the interactions between species, not just external factors like the weather. Around a decade ago, Nelson Hairston's team at Cornell University in New York began experimenting on single-celled algae and tiny animals called rotifers that feed on them. They expected to see a classic predator–prey cycle – a decline in algae as rotifers increased, followed by a rotifer crash as they ran out of food, leading the algae to rebound, and so on.

In fact, they saw unexpected patterns. Sometimes rotifer numbers grew even when algal numbers remained constant.

The reason, Hairston realized, was that these algae were evolving rapidly, alternating between putting resources into defence or into multiplying – which creates more rotifer food. Rotifer numbers sometimes increased at just the right rate to keep the rapidly reproducing algae in check. When the team repeated the experiments with genetically identical algal cells, to slow evolution to a crawl, they saw classic cycles.

Hairston later discovered that theoretical biologists had predicted that rapid evolution could produce the kinds of patterns he saw. What remains unclear is how common it is in the wild.

One place this kind of cycle might be getting under way is the Hawaiian island of Kauai, where the crickets recently fell silent. In the 1990s, a parasitic fly arrived which tracks down male crickets calling for mates and deposits its egg on them. The larvae then devour the crickets alive. The cricket population plummeted.

In 2003, the island was still silent – so Marlene Zuk of the University of Minnesota was surprised to find plenty of crickets there. It turned out that almost the entire population had a mutation that alters the wings of male crickets and prevents them making any sound when rubbed together. The population has survived because a few males can still chirp. Silent males gather around these males and intercept potential mates.

For Zuk, the interesting question is what will happen next. At present, the crickets are heading down an evolutionary dead end. 'I don't think that a completely silent population could survive,' says Zuk. Instead, she suggests, we will see a predator–prey cycle driven by rapid evolution, similar to the ones Hairston observed. As silent males increase, parasite numbers may fall, leading to a rebound in singing males followed by a parasite revival, and so on.

## Switching direction

There is nothing new about the idea of an evolutionary arms race in which species have to continually evolve to keep up – it is called the Red Queen hypothesis. What is new, though, is the idea that not only can this kind of evolution occur far more rapidly than once thought, but that the runners in the race keep switching direction.

Put it all together and the picture of evolution that is emerging is radically different to the way most people envisage the process. As Kinnison puts it, the popular view of evolution is upside down. People think evolutionary changes are imperceptible in the short term but add up to big changes over millions of years. In fact, the opposite is true. It now appears that organisms evolve very rapidly in response to any changes in their environment, but in the longer term most evolutionary changes cancel each other out. So the longer the period you look at, the slower evolution appears.

---

### Rapid evolution in humans

Members of the Fore tribe of Papua New Guinea used to believe that when someone died, their loved ones should eat every bit of the body. The daughters ate the brain and sometimes fed titbits to their children. This tradition led to the spread of a degenerative brain disease called kuru. Like Creutzfeldt-Jakob disease, it is caused by a rogue prion protein that accumulates in the brain.

Kuru killed nearly all the young women in some villages. But a few did not succumb. They were the descendants of a person born around 200 years ago with an unusual mutation in the prion protein that stops it going

---

rogue. As kuru became widespread, the mutation rapidly became more common. Half of the women in the areas most affected now carry the mutation, which has not been found anywhere else in the world. If the tradition of cannibalism had not been stopped in the 1950s, it would have become even more common among the Fore.

The emergence of kuru resistance is one of the clearest examples of very rapid human evolution but it is far from the only one. Around 3000 years ago, the ancestors of Tibetans split from the population that gave rise to the Han people of China. As soon as they began living at altitude, the population began to adapt. While some of the adaptations are a result of living in the mountains – a bit like altitude training in athletes – some are genetic.

One variant in a gene controlling the production of red blood cells, for instance, is found in 78 per cent of Tibetans but just 9 per cent of Han people. This selection process is thought to be ongoing, according to the authors of this study.

More evidence comes from a study of Tibetan women living above 4000 metres. Those with high levels of oxygen in the blood had 3.6 surviving children on average, whereas those with low oxygen levels had just 1.6, due to much higher infant mortality. That suggests the genetic variant thought to be responsible for higher blood oxygen levels is being passed on in greater numbers and becoming more common.

## How new species are formed

Not long ago, we thought we knew how species formed. We believed that the process almost always started with complete isolation of populations. It often occurred after a population had gone through a severe 'genetic bottleneck', as might happen after a pregnant female was swept off to a remote island and her offspring mated with each other.

The beauty of this so-called 'founder effect' model was that it could be tested in the lab. In reality, it just didn't hold up. Despite evolutionary biologists' best efforts, nobody has even got close to creating a new species from a founder population. What's more, as far as we know, no new species has formed as a result of humans releasing small numbers of organisms into alien environments.

These days, the focus has changed. Biologists still think that most speciation is 'allopatric' – the result of geographical isolation – but the ideas have shifted away from chance and small populations. Biologists are looking instead at all kinds of weird ways that species can change rapidly. The main forces at work are ecological selection (where new species form as the result of adaptations to changing environmental conditions) and sexual selection (in which changing sexual traits and preferences for such traits lead to divergence in populations). The big questions revolve around the relative importance of these two forces.

One of the most dramatic illustrations of the power of ecological selection is 'parallel speciation', where essentially the same species arise independently in different places in response to a similar environment. The best example of this is among stickleback fish in Canadian lakes. Several lakes contain two different species of stickleback, one bottom-feeding and the other plankton-eating. Analysis of mitochondrial DNA (mtDNA) reveals that sticklebacks living in the same lake are more closely related to each other than to their counterparts in different lakes. In other words, they probably arose by parallel speciation.

These findings also indicate 'sympatric speciation' – speciation without geographical isolation. Biologists who believe a species almost never splits into two without first being physically isolated have hotly contested the idea of sympatric speciation. But those who support sympatric speciation have seized on the stickleback findings, as well as mtDNA studies of several other species, which seem to support the idea.

Sympatric speciation remains contentious, but other research suggests how it might occur. The evidence comes from a group of fish that have undergone the most extravagant burst of speciation we know of: the cichlids of Africa's great lakes. Between them, Lakes Malawi, Victoria and Tanganyika contain around 1700 species of cichlids, many of which have evolved since the last ice age, a mere 12,500 years ago. One puzzle about cichlids was explaining the evolution of over 500 species in Lake Victoria that live together without physical barriers to prevent interbreeding. Sexual selection seems to be the key, with males varying in colour and picky females showing distinct colour preferences. In this way, populations of fishes that looked remarkably similar in every other way might have become reproductively isolated, with sexual selection leading eventually to the emergence of new species.

This particular form of sexual selection relies on females being able to distinguish between differently coloured males. But as pollution clouds the waters of Africa's great lakes, cichlids are losing this ability. In the murky waters, hybridization is becoming increasingly common, and because cichlid species are evolutionarily close, they often produce viable hybrid offspring. Surprisingly, some biologists now think that hybridization might actually be a creative process, churning out new species, and it has probably happened naturally in Lake Victoria many times in the past. Hybridization may be a significant factor in some of the evolutionary explosions we call adaptive radiations.

In theory, we can test whether species are the product of parallel evolution, sexual selection or hybridization by looking for 'speciation genes' – genes that are responsible for preventing interbreeding. However, this line of research has not turned up many major-effect speciation genes. Instead it seems that selection often operates on genetic variation caused by multiple genes of small effect. Populations of organisms often appear to contain the variation needed for dramatic adaptive change. This means that evolution doesn't need to wait around for mutations but can get cracking as soon as the ecological circumstances warrant it. This really wasn't what twentieth-century evolutionary biologists would have predicted.

Speciation is still a lively area of study, and the falling costs of genome sequencing mean that we are now entering an exciting era of population genomic studies of species formation.

## Evolvability – the capacity of evolutionary innovation

The remarkable diversity of life on Earth stands as grand testimony to the creativity of evolution. Over the course of 500 million years, natural selection has fashioned wings for flight, fins for swimming, and legs for walking, and that's just among the vertebrates. The capacity for evolutionary innovation – or, in buzzword form, 'evolvability' – is built into the fabric of life.

Few questions are more fundamental to evolutionary theory, yet evolvability didn't enter biological parlance until 1987, when arch-phrase-maker Richard Dawkins coined the term. In the intervening decades it has become a hot topic, though only recently has real-world evidence begun to put flesh on the bones of the theory.

Now there are studies aplenty to shed light on the factors that might constrain and enhance an organism's capacity to evolve. They are also explaining crucial events in human evolution, such as the switch to walking on two legs and the emergence of our highly dexterous, tool-using hands.

One of the first hurdles has been to define exactly what 'evolvability' means. The aim is to capture the capacity of a species or population to respond to natural selection. Since genetic variation is the raw material on which selection acts, the extent of this variation in a population provides a crude measure of evolvability.

When most researchers talk about evolvability, however, they mean something more subtle – not just how much genetic variation is present, but whether this variation translates into adaptive changes in the organism's outward appearance and behaviour that could be shaped by natural selection. Günter Wagner, a pioneer in the field who is based at Yale University, therefore defines evolvability as 'the capacity to generate heritable phenotypic variation'. That is to say, variation in an organism's body-type that can be passed from generation to generation.

## Robustness

The real question, of course, is what determines this capacity. Two factors are key. Perhaps the most fundamental of these is an organism's 'mutational robustness' – the capacity to develop normally despite the presence of genetic mutations. Since genes rarely act independently, a particular mutation could have a positive, negative or neutral effect on the organism, depending on the overall genetic background. Greater robustness could therefore be achieved by mechanisms that dampen the impact of mutations in any particular gene. In principle, that should increase an organism's survival because it reduces the chance of potentially harmful changes to the organism's body plan. But this buffering effect would also be the enemy of change, masking potentially beneficial variation and putting the brakes on the organism's evolvability.

Or so it might seem. In fact, by neutralizing the effects of otherwise harmful mutations, robustness preserves genetic variations that might otherwise be weeded out. That means organisms accumulate a wealth of hidden mutations within the population. Further genetic or environmental changes might then remove the buffering mechanisms and unmask the effects of these stored mutations, providing ready-made variation in the organism's make-up. What might the mechanisms behind this robustness be?

The main players seem to be 'heat-shock proteins', according to studies led by Susan Lindquist at the Massachusetts Institute of Technology. HSPs ensure that other proteins always fold in the same stable three-dimensional shape, which is vital for their role within the cell. Under harsh conditions like scorching temperatures or high salinity, proteins can fold wrongly, preventing them from carrying out their function. It is here that the HSPs step in, acting as chaperones to guide the protein into the correct shape and allowing it to function properly even under troublesome circumstances.

Crucially, HSPs also ensure a protein folds into the same stable shape even in the face of genetic mutations that shuffle the protein's sequence of amino acids. That permits hidden variations to build up over time without getting in the way of the protein's everyday activities.

The structure and function of proteins govern all kinds of processes in the development of an organism. So when

Lindquist's team knocked out the HSPs in thale cress and fruit flies, the stored mutations suddenly became apparent in physical changes to the organisms, including new leaf shapes for the cress and changes to the shape of the flies' eyeballs. The genes that code for HSPs don't routinely stop functioning in natural populations but occasionally changes to the environment, like a radical change in diet, can overwhelm the HSP system and lead to similar effects, providing variation that can respond to the new evolutionary pressures precisely when it is needed most.

HSPs are not the be-all and end-all of robustness. Some proteins are intrinsically more robust than others, even without the help of HSPs. This too can affect an organism's evolution. In 2006, for example, Jesse Bloom at the Fred Hutchinson Cancer Research Center, Seattle, and colleagues, showed that more robust proteins can pick up useful additional functions from new mutations without losing their basic structure and collapsing into a useless tangle. And in 2014, Bloom demonstrated that the ability of the influenza virus to tolerate mutations enables it to adapt to the pressure of immune system attack.

Other research led by Robert McBride, then at Yale University, showed that viruses bred to produce more robust proteins adapted to new evolutionary pressures, like higher temperatures, more quickly than less robust strains. In other words, they were more evolvable.

## Integration

Robustness is only half of the story of evolvability, however. The second key factor concerns a phenomenon known as integration – the way different body parts or traits appear to vary and evolve together. Integration between traits often results from their shared evolutionary history. Body parts like limbs, teeth, ribs and vertebrae that are repeated along the body axis, for example, arose through the direct duplication of certain genes way back in evolutionary history. The two copies will not be completely independent of one another because the expression of both will ultimately be governed by the same regulatory genes at a different point in the genome, meaning the two body parts would still tend to vary and evolve together.

Integration is also likely if the different parts are involved in the same function. The four fingers and thumb, for example, work together for grasping and manipulation. To preserve the optimal use of the hand, changes in one part – a longer finger, say – need to be complemented by corresponding changes in the other digits. Selection therefore favours a developmental system in which genetic changes affecting digit length produce a coordinated shift in all digits.

Like robustness, integration can be a double-edged sword. On one hand, it increases the capacity to generate coordinated, adaptive changes in body structures, which surely help to increase an individual's chances of survival. On the other, it also constrains the possible evolutionary avenues an animal could take – capping its evolvability, in other words – since a potentially beneficial change to one trait could have a disastrous impact on the other traits to which it is linked.

Luckily for life on Earth, integration is not an all-or-nothing affair. It is becoming clear that the traits can be integrated to varying degrees (see 'Evolvable dogs'). Sometimes existing integration can be uncoupled altogether, making each trait an independent 'module' that is more evolvable.

Consider the evolution of wings in mammals. The fore and hind limbs of mice and other rodents are very tightly integrated, so that changes in one limb pair, such as an increase in length, correlate almost perfectly with changes in the other pair.

FIGURE 7.2 The forelimbs and hind limbs of bats have evolved to perform different functions, unlike those of rodents.

Bats, however, use their modified forelimbs for flying and their hind limbs for grasping – two very different tasks, which would hint that the limbs are quite loosely integrated.

## Arms, legs and wings

Sure enough, Benedikt Hallgrimsson of the University of Calgary, Canada, and Nathan Young of the University of California, San Francisco, found that the co-variation in the length of bones in the bat's fore and hind limbs tends to be much lower than in other mammals. The implication is that the ancestors of bats must have lost the genetic integration between fore and hind limbs somewhere along the evolutionary path, opening the door for the evolution of wings.

A similar story could explain various stages of primate evolution, too. Campbell Rolian at the University of Calgary, for example, has compared quadrupedal primate species such as macaques, in which the hands and feet share similar roles, with species like the great apes (humans, chimpanzees, gorillas and orangutans), in which the hands and feet perform independent functions.

As expected, Rolian found greater integration between the hands and feet of quadrupeds than in the great apes. Another analysis of the entire fore and hind limbs, rather than the individual hands and feet, found a similar result, with roughly 40 per cent less co-variation in the ape limbs compared with the quadrupeds.

The result was that arms and legs could respond to natural selection with a greater degree of independence, thereby increasing their evolvability. Ultimately, this enabled early humans to evolve longer legs adapted to walking and running while leaving arm length relatively unchanged. By the same logic, a shortening of the forearm, which would have facilitated tool use, was not constrained by corresponding changes on the lower leg that might have reduced their walking power.

Importantly, integration comes in degrees. Even though the integration between the arms and legs has been significantly weakened over the years, some developmental connections linger that are strong enough to have changed the course of evolution in intriguing ways. For example, Rolian and Hallgrimsson, working with Daniel Lieberman of Harvard University, discovered that evolutionary pressures on the feet might have prepared humankind's highly dexterous hands for tool use and manual tasks. The team's discovery came from a detailed comparison of the length of the bones making up each toe and corresponding finger. They found that sufficient integration remained between the fingers and toes for them to have co-evolved to a certain degree.

Evolutionary pressures shaping the feet could therefore have changed the hands, or vice versa – but which way round? Using computer simulations to estimate the possible evolutionary pressures and corresponding changes to primate anatomy, the team suggested that natural selection acted primarily on the toes, enlarging the big toe and reducing the outer digits to stabilize the feet for walking.

The result, of course, was that as the big toe evolved, the thumb also grew accordingly. Completely by chance, that meant that the tips of the thumb and the fingers could now meet each other for the first time in evolutionary history, providing our ancestors with greater dexterity and precision gripping which became the key to the successful use of tools.

## Evolvability through the ages

The take-home message is that animals are made up of a 'nested hierarchy' of modules and integrated traits. So while the bones of the arms and hands show reduced integration in humans (and great apes generally) compared with quadrupedal monkeys, integration was nonetheless strong enough between human hands and feet to have profound evolutionary consequences. It is these specific patterns of integration and modularity, rather than either factor independently, that ultimately determines evolvability.

Indeed, looking deeper into prehistory, it is easy to see how these factors could have played a crucial role throughout animal evolution. About 540 million years ago, the Cambrian explosion led to the emergence of the basic body plans of the 35 or so phyla of animals recognized today. Their common ancestor had not achieved a high level of integration or robustness, making it developmentally flexible and primed for evolutionary innovation. Evolution cashed in on that flexibility but soon pushed for greater developmental integration, more or less fixing the 35 body plans in the process.

That's not to say evolvability has taken a nosedive since then. Although the developmental processes that produce the basic body plans of animals are too tightly integrated for fundamental change, a further drive towards greater modularity among parts of animals has increased their individual evolvability. It is this tinkering with bits and pieces of animal bodies, rather than radical reinvention of the body, that has fuelled the astonishing biological innovation among arthropods and vertebrates in particular.

Finding out exactly what triggers the dissociation of integrated traits, prompting them to become increasingly modular, is now a key goal. It is possible that, in some cases, body parts become dissociated as a lucky accident that evolution then capitalizes on. Getting a better handle on the process of dissociation requires a deeper understanding of the genetic mechanisms involved. Steps are already being made in this direction by mapping genes that determine the integration among various characteristics in mice. But the genetic detective work is revealing a complex picture. For example, a huge number of gene variants are implicated in determining the facial shape of humans, says Hallgrimsson. 'You can't explain a lot of the variation from the genomic data. So many complex traits are turning out to be like that.' Changes in the way genes are regulated, rather than changes in the genes themselves, is key. Intriguingly, other research has found that evolvability itself is evolvable.

These are early days in the empirical study of evolvability. Further progress will depend on researchers drawing together data on diverse aspects of biology. A complete theory of integration, modularity and the developmental basis of evolvability will require connecting genetics and developmental biology with morphological studies in both experimental and natural settings.

## Evolvable dogs

The faces of man's best friend come in an astonishing range – from the short, squashed face of a Pekinese to the slender snout of a collie. Studies by Abby Drake, then at the University of Manchester in the UK, and her colleague Christian Klingenberg, have shown that diversity in dog faces is comparable to the diversity apparent between all carnivore species.

But the remarkable variation in dogs has been achieved in just a few thousand years of selective breeding, and was possible thanks to a limited integration between the face and brain – a lack not normally found in other mammals. Intriguingly, this modularity is also found in wolves, coyotes and jackals, suggesting dog faces were always this evolvable – they simply needed the right evolutionary pressures to shape their snouts.

FIGURE 7.3 The faces of dogs are remarkably evolvable.

## Testing evolution in the lab

Evolutionary experiments in the lab are now becoming routine, but the longest running one, kick-started in 1988, has allowed us to witness evolution in unprecedented detail. It has shown how a major change in one creature can transform its environment, and alter the evolutionary trajectory of all the creatures inhabiting that space.

The Long-term Experimental Evolution Project was set up by Richard Lenski at Michigan State University, who took a single strain of the E. coli bacterium and set up 12 cultures.

Every day since then, a sample of each culture has been transferred to a fresh growth medium, containing glucose as the main nutrient. The bacteria have now undergone more than 66,000 generations since the experiment began. Samples are frozen every 75 days creating an artificial 'fossil record' so the team can go back and identify the precise genetic mutations underlying the changes they see.

The biggest evolutionary shift occurred after about the 31,500th generation, when one line in one of the 12 populations evolved the ability to feed on citrate, another chemical in the growth medium. E. coli don't normally feed on citrate because they can't carry it into their cells. But a mutation in the citrate-eaters enabled them to make an 'antiporter' protein, CitT, that allows citrate to cross the membrane and enter the cell. The gene for this protein already existed, but it's usually switched off when oxygen is present.

The antiporter is a kind of revolving door. It allows one molecule to be swapped for another. In this case, the citrate is imported into the cell in exchange for one of three smaller, less-valuable molecules: succinate, fumarate or malate. Once this ability to feed on citrate evolved the population boomed because the same growth medium could now sustain more cells.

Those citrate feeders soon became dominant, outcompeting all but one other strain of E. coli, which in turn evolved to exploit the changed environment – which now contained the three exported molecules. It did this by making more of a transporter protein called DctA, which imports – at a small energy cost – succinate and other molecules exported by the citrate-eating strain.

But things did not stop there. The citrate-eaters then also started making more DctA to try to claw back some of the succinate and other molecules they were losing in the process of acquiring citrate.

The work is a neat example of how evolution and ecosystems are inextricably linked, showing how evolutionary novelties can change environmental conditions, thereby facilitating diversity and altering both the structure of an ecosystem and the evolutionary trajectories of coexisting organisms.

The researchers compare this to the evolution of photosynthetic bacteria some 2.4 billion years ago: just as oxygen excreted by the first photosynthesizers transformed Earth and changed the course of evolution, so the appearance of citrate-eaters altered the growth medium and changed the evolutionary path of all the bacteria living in it.

These findings are also yet another example of the mindlessness of evolution. The best solution would be to use a little energy to import citrate directly, rather than swapping it for succinate and then spending energy to try to get that succinate back before other bacteria can feed on it.

The experiment also shows that, when it comes to evolution, there is no such thing as perfection. Even in the simple, unchanging environment of a laboratory flask, bacteria never stop making small tweaks to improve their fitness.

## No upper limit

After 10,000 generations, Lenski thought that the bacteria might approach an upper limit in fitness beyond which no further improvement was possible. But 50,000 generations of data showed that wasn't the case. When pitted against each other in an equal race, new generations always grew faster than older ones. In other words, fitness never stopped increasing.

Their results fit a mathematical pattern known as a power law, in which something can increase forever, but at a steadily diminishing rate. 'Even if we extrapolate it to 2.5 billion generations, there's no obvious reason to think there's an upper limit,' says Lenski.

Lenski's results suggest that evolution never reaches a pinnacle of perfection where progress stops, even in the simplest and most constant environments.

That undermines one of the favourite metaphors of evolutionary biology, which sees species evolving towards peaks of fitness in a landscape of possibilities. In the real world, species live in ever-changing environments. The results show that there are even more ways of adapting to environments than we imagined.

### Steady or flashy: Who wins in evolution death match?

One Petri dish. Two bacterial strains competing to take it over. Given 500 to 1500 generations to evolve, which will prevail?

By staging survival battles that pitted clones of Escherichia coli against one another, researchers have shown that the ultimate victors are seldom the early pacesetters, who owe their success to specific gene mutations. Instead, it's the 'plodders' that eventually prevail, mainly because, unlike the early front runners, they remain able to acquire the modest but valuable mutations that are ultimately vital for survival and domination.

These less dramatic mutations give them the edge in the end because they play to the strengths of the underlying genome as a whole. The 'flashy' mutations that confer dramatic early success don't mesh as well with the entire genome as the slower-emerging ones.

The battles were staged by a team led by Richard Lenski and Jeffrey Barrick of the University of Texas at Austin. They selected four distinct pairs of clones and pitted each pair against one another. The front-runner strains, which began to take over the Petri dish early on, developed beneficial mutations improving efficiency of topA, a gene that transcribes many other genes. They also developed mutations in rbs, a gene that improves production of DNA and RNA.

But in the end, as in the fable of the tortoise and the hare, these mutations proved to be their downfall because they ended up having an evolutionary 'sleep': their plodding adversaries carried on evolving much less dramatic but ultimately pivotal mutations that gave them the edge.

This study helps to answer questions about whether selection happens mainly through dramatic mutations in single genes or steady evolution of whole genomes. The discovery of the evolutionary pre-eminence of the whole genome over the effect of individual genes in bacteria has implications for human genetic studies. It may help explain why gene researchers often struggle in vain to try to identify individual genes linked with strongly heritable traits.

# 8
# Evolving questions

*Although the basic concepts are firmly established, there are many issues with our understanding of evolution that are still live – and often controversial. Should the idea that evolution is directionless be revised? What role does epigenetics play in evolution? Can organisms adapt first, then mutate later? And is evolution predictable?*

## Is it time to put progress back in the picture?

The concept of progress has been purged from evolutionary theory, but could it be time to let it back in?

The celebrated palaeontologist Stephen Jay Gould once wondered what would happen if we could rewind the tape of life. If it were possible to turn the clock back half a billion years and then let evolution happen all over again, what would we see? Gould famously argued that the history of life would not repeat itself. The world would be unfamiliar, and would probably lack humans.

His point was to demonstrate that evolution is not a process of inexorable progress but of contingency. Mutations happen unpredictably. Sexual reproduction combines genes at random. Droughts, ice ages and meteorites strike without warning and kill off fully fit individuals and species.

We tell ourselves stories of evolutionary progress but these are just wishful thinking. Life produces abundant variations; most fail. The few that survive we call the most advanced, but that is a profound error which conflates 'latest' with 'best'. As Gould wrote in his classic book *Wonderful Life* (1989): 'Life is a copiously branching bush, continually pruned by the grim reaper of extinction, not a ladder of predictable progress.'

Gould also had little time for humanity's hubris. Far from being the pinnacle of evolution, we are just another product of contingency. 'Perhaps,' he wrote frostily, 'we are only an afterthought, a kind of cosmic accident, just one bauble on the Christmas tree of evolution.'

Gould's view is the orthodoxy of evolutionary theory. Yet it remains hard to reconcile with the intuitive sense that life has indeed progressed over time. All life was once single-celled, yet now a single organism can contain tens of trillions of cells. The number of cell types has increased, too, from one kind in single-celled organisms to 120 in mammals. Brains have grown larger. And humans have accelerated this trend in the past 50,000 years with our own uneven but powerful ascent.

For many years, a small but energetic group of researchers has been trying to rehabilitate the concept of evolutionary progress and explain it in theoretical terms. They hope to show that Gould's view of evolution is too bleak and that certain kinds of biological progress are not merely accidental or illusory, but necessitated by physical law. If these researchers succeed, it could lead to a crucial modification of current theory.

Gould and those who followed in his footsteps accepted that life has increased in size, complexity and diversity. However, they argue that this is not because evolution is inherently progressive.

Instead, it is an illusion. By definition the first life was very simple. As variation increased, some organisms inevitably became more complex. Humans pay the most attention to the complex ones, leading to a belief in an upward march. As Sean B. Carroll, a professor of molecular biology at the University of Wisconsin–Madison, puts it: when there is nowhere to go but up, some species will go up.

Development-oriented theorists accept these passive increases in complexity. But they argue that there are also 'driven' processes that bias evolution toward increasing complexity. John Smart, a member of the evolution, complexity, and cognition research group at the Free University of Brussels in Belgium and a leading thinker in this field, argues that evolution and development can be reconciled. That is, it will be possible to define progress in objective terms and explain why it must happen. The case laid out by Smart and other theorists is based on at least four arguments.

The first concerns a new way of thinking about progress – a concept that is notoriously difficult to define, largely because what counts as progress depends on who is doing the defining. More complexity seems valuable to us, for example, but many organisms – especially parasites – are successful thanks to a reduction in complexity.

## Energy flows

Rooting a new definition in basic physics would be one way around this problem.

Eric Chaisson, an astrophysicist at the Harvard-Smithsonian Center for Astrophysics, Cambridge, Massachusetts, has put forward the idea of energy rate density, a measure of how much energy flows through each gram of a system per second. A star, for all its spectacular output, has a much lower energy rate density (2 ergs per gram per second) than a houseplant (3000 to 6000 ergs per gram per second). This sounds counter-intuitive until you remember that stars are just balls of gas.

Humans do better still, with a basic energy rate density of 20,000 ergs per gram per second. Societies, too, can be measured in this way. Chaisson estimates that hunter-gatherer societies have an average energy rate density of 40,000 ergs per gram per second, while technological societies use 2 million ergs per gram per second.

Chaisson argues that energy rate density is a universal measure of the complexity of all ordered systems, from planets and stars to animals and societies. Furthermore, when he plots the energy rate density of such ordered systems against the time they first appear in the history of the universe, the line goes

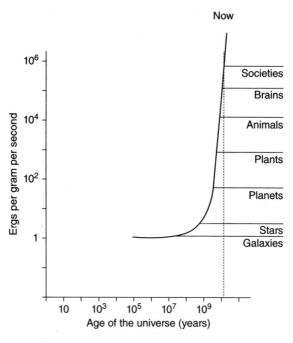

FIGURE 8.1  Growing energy density: as the universe ages, ever more complex systems evolve.

unequivocally upwards, indicating a general increase in complexity over time (see Figure 8.1).

## Thermodynamics

The second argument concerns thermodynamics. At first sight, the second law of thermodynamics is a gloomy affair. It seems to indicate that increases in disorder are inevitable and irreversible and that the universe is running out of the energy needed to create and sustain complex entities such as living things.

A literal reading of this law implies that the ascent of life is extremely unlikely. More nuanced readings, however, have been used to argue that local increases in complexity are not merely permitted by the law, but required by it, and order can and does emerge spontaneously from chaos.

Physicist J. Miguel Rubí of the University of Barcelona in Spain says that, strictly speaking, the second law of thermo-dynamics applies only to systems in equilibrium, a state in which nothing changes. This condition is rarely present in the universe. Earth, for example, is heated by the sun, creating energy gradients on its surface. Where energy gradients exist, pockets of complexity can arise even as the system as a whole decays into disorder. These pockets provide a foothold for fur-ther increases in complexity. Energy gradients thus provide a loophole in the second law that permits life to arise and ascend.

## Convergent evolution

Argument number three is convergent evolution. Taking a dif-ferent view to Gould's argument, the tape of life has been rerun many times – at least partially. In many cases, very different spe-cies living in similar environments have independently evolved in similar ways.

In his book *What Technology Wants* (Viking Press, 2010), Kevin Kelly, the founding editor of *Wired* magazine, gives numerous examples of convergent evolution to support his argument that the outcomes of evolution – of which he considers technology one – are not accidental. Flapping wings evolved independently in birds, bats and pterodactyls. Dolphins, bats and several species of cave-dwelling bird separately hit on echolocation. Fish in the Arctic and the Antarctic independently evolved antifreeze compounds. Perhaps the best-known example is the camera eye, which has evolved independently at least six times. The implication, Kelly writes, is that many outcomes of evolution are not accidental but inevitable. These outcomes include not just organs but brains, minds, societies, and technologies.

Intelligence may be another convergent property. Nicola Clayton, a professor of comparative cognition at the University of Cambridge, and Nathan Emery, a cognitive biologist at Queen Mary, University of London, argue that while primates and crows are far apart on the evolutionary tree and have very different brain structures, they have independently evolved many similar kinds of cognition, including tool use, deception and complex social groupings. The implication, again, is that intelligence always emerges in favourable conditions.

## Catastrophes

Last of all, a theory of development must account for catastrophism. The occurrence of unpredictable, planet-altering events is a challenge for any developmental perspective on evolution. Had the dinosaurs not been killed by an impact, critics of the theory say, mammals would not have had the opportunity to expand into new niches and there would have been no evolutionary sequence leading from primates to tool-wielding, language-using apes. In short: no impact, no us.

Simon Conway Morris, a palaeontologist at the University of Cambridge, counters this by arguing that while catastrophes delay or accelerate the developmental process they do not significantly change it. The key is convergent evolution.

Suppose the deadly meteorite had sailed harmlessly by, Conway Morris suggests. The dinosaurs would have survived for the next 30 million years until Earth's next glaciation. The cold would have killed off those dinosaurs living north and south of the tropics, opening up niches for the warm-blooded mammals and birds that co-existed with them. Eventually tool-users not unlike us would have evolved and sooner or later any dinosaurs remaining in the tropics would have been hunted to extinction. 'The mass extinction of the dinosaurs would then have been under way, perhaps 30 million years behind schedule in comparison with the real world,' Conway Morris writes.

So whatever catastrophe hits, the tape of life would probably run more or less the same way. It might delay the developmental process by reversing an advance, but the advance would eventually happen again. Or it might accelerate the process by opening up an environmental niche. In either case, the outcome would not change substantially, only the timing would.

If these four arguments hold up, it would mean a significant expansion of evolutionary theory is needed, showing that life not only evolves, it develops.

The implications would be profound. Development, unlike evolution, has a direction: an acorn becomes a tree, an embryo becomes a newborn. It never goes the other way. And while the outcome is not fully determined, it is powerfully constrained.

Direction and constraint, however, do not imply design and purpose. A developmental view of evolution needs no help from teleology. Such a theory of evolution offers no support for intelligent design. Indeed, it would strike another major blow to it by offering a cogent naturalistic explanation for the emergence of complexity.

Perhaps more profoundly, admitting progress into evolution would give a different perspective on our own existence. In offering a naturalistic explanation for the emergence of intelligence and its offspring, language and technology, it would cast them as predictable outcomes of the cosmos rather than as accidents of contingency. Far from being 'just one bauble', we would have an explicable, even inevitable, place in the order of things.

## Adapt first, mutate later: is evolution out of order?

We used to think evolution had to start with random mutations; now walking fish and bipedal rats are turning our ideas on their head. We have long known that our muscles, sinews and bones adapt to cope with whatever we make them do. A growing number of biologists think this kind of plasticity may also play a key role in evolution. Instead of mutating first and adapting later, they argue, animals often adapt first and mutate later. This process could even play a role in major evolutionary transitions such as fish taking to land and apes starting to walk upright.

The idea that plasticity plays a role in evolution goes back more than a century. Some early biologists thought that characteristics acquired during an animal's lifetime could be inherited by their offspring: giraffes got their long necks by stretching to eat leaves, and so on. The French naturalist Jean–Baptiste Lamarck is the best-known advocate of this idea, but Darwin believed something similar. He even proposed an elaborate mechanism to explain how information about changes in the body could reach eggs and sperm, and therefore be passed on to offspring. In this way, Darwin suggested, plasticity produces the heritable variations on which natural selection can work its magic.

With the rise of modern genetics, such notions were dismissed. It became clear that there is no way for information about what animals do during their lifetime to be passed on to their offspring (although a few exceptions have emerged since). And it was thought this meant plasticity had no role in evolution.

Instead, the focus shifted to mutations. By the 1940s, the standard thinking was that animals mutate first and adapt later. A mutation in a sperm cell, say, might produce a physical change in the bodies of some offspring. If the change is

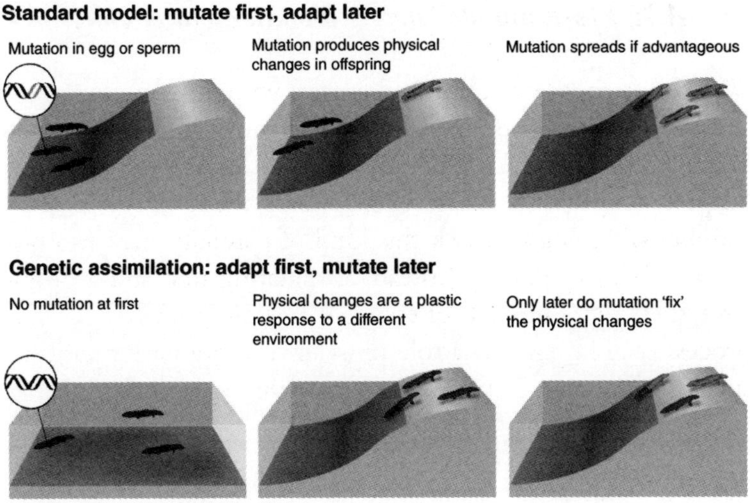

**Standard model: mutate first, adapt later**

Mutation in egg or sperm

Mutation produces physical changes in offspring

Mutation spreads if advantageous

**Genetic assimilation: adapt first, mutate later**

No mutation at first

Physical changes are a plastic response to a different environment

Only later do mutation 'fix' the physical changes

FIGURE 8.2 Evolving without evolving.

beneficial, the mutation will spread through the population. In other words, random genetic mutations generate the variation on which natural selection acts. This remains the dominant view of evolution today.

The dramatic effects of plasticity were not entirely ignored. In the 1940s, for instance, the Dutch zoologist Everhard Johannes Slijper studied a goat that had been born without forelegs and learned to hop around, kangaroo-like, on its rear legs. When Slijper examined the goat after its death, he discovered that the shape of its muscles and skeleton looked more like those of a biped than a quadruped.

Few biologists considered such findings relevant to the evolutionary process. The fact that changes acquired during an animal's lifetime are transient seemed to rule out that possibility.

## Transient response

But what if the environmental conditions that induce the plastic response are themselves permanent? In the wild, this could happen as a result of alterations in prey animals, or in the climate, for instance. Then all the members of a population would develop in the same, consistent way down the generations. It would look as if the population had evolved in response to an altered environment, but technically it's not evolution because there is no heritable change. The thing is, the only way to tell would be to 'test' individuals by raising them in different circumstances.

In this way at least, plasticity can allow animals to 'evolve' without evolving. The crucial question, of course, is whether it can lead to actual evolution, in the sense of heritable changes. The answer, surprisingly, seems to be yes. In the 1950s, British biologist Conrad Hal Waddington showed that it is feasible in an experiment involving fruit flies. Waddington found that when pupa are briefly heated, some offspring develop without crossveins in their wings. He then selected and bred those flies. By the 14th generation, some lacked crossveins even when their pupa were not heated. A physical feature that began as a plastic response to an environmental trigger had become a hereditary feature.

How is this possible? Plastic changes occur because an environmental trigger affects a developmental pathway in some way. More of a certain hormone may be produced, or produced at a different time, or genes are switched on that normally remain inactive, and so on. The thing is, random mutations can also have similar effects. So in an environment in which a particular plastic response is crucial for survival, only mutations that reinforce this response, or at least do not impede it, can spread through a population. Eventually, the altered developmental pathway will become so firmly stabilized by a genetic scaffolding that it will occur even without the environmental trigger, making it a permanent hereditary feature.

## Genetic assimilation

Waddington called this process genetic assimilation. It may sound like Lamarckism, but it is not. The acquired characteristics don't shape the genetic changes directly as Darwin proposed, they merely allow animals to thrive in environments that favour certain mutations when they occur by chance.

Waddington's findings have been regarded as a curiosity rather than a crucial insight. But in the past decade or two, attitudes have begun to change. One reason for this is a growing appreciation of the flexibility of genes. Rather than being rigidly preprogrammed, we now know that the environment influences many aspects of animals' bodies and behaviour.

Such discoveries have led some biologists to claim that developmental plasticity plays a major role in evolution. A few, such as Kevin Laland at the University of St Andrews, UK, even argue that the conventional 'mutate first, adapt later' picture of evolution needs a rethink (see Chapter 11). Most biologists have yet to be convinced.

The sceptics point out that genetic assimilation does not overturn any fundamental principles of evolution – in the long run, evolution is all about the spread of mutations, whether or not plasticity is involved. Yes, say the proponents of plasticity, but the key point is that plasticity can determine which mutations spread, so its role should be given the prominence it deserves.

All this still leaves open the question of whether genetic assimilation can 'fix' traits that first appear as a result of plasticity. A decade ago, Richard Palmer at the University of Alberta in Edmonton, Canada, found a way to search for evidence in the fossil record. Most animals have some asymmetric traits. In our case, it's the position of the heart and other organs, which is encoded in our genes. But in other species, asymmetries are plastic. For instance, the enlarged claw of male fiddler crabs is as likely to be on the left as on the right.

What Palmer showed by examining the fossil record of asymmetry in 68 plant and animal species is that on 28 occasions, asymmetries that are now hereditary and appear only on one side started out as non-hereditary asymmetries that appeared on either side. 'I think it's one of the clearest demonstrations that genetic assimilation has happened and that it is more common than expected,' says Palmer.

There is a caveat here, though. The ancestral non-hereditary asymmetries may have been a result of random genetic noise, says Palmer. So while his work does show genetic assimilation in action, it was not necessarily fixing traits due to developmental plasticity.

## What is the role of epigenetics in evolution?

*The term 'epigenetics' refers to an array of molecular mechanisms that affect the activity of genes. Epigenetic 'switches' turn gene activity up or down. They have long-lasting effects that can persist through cell division, and sometimes through sexual reproduction too. Here, geneticist Adrian Bird examines the evidence that epigenetic traits can be passed down through the generations.*

We started to get the first glimpse of some of the mechanisms involved in epigenetic phenomena in the 1970s and 1980s. The first to be discovered was DNA methylation, which involves a small chemical subunit called a methyl group being added to DNA. Other epigenetic mechanisms include chemical changes to the proteins that package DNA. One of the most interesting aspects of epigenetics is the route it offers for the environment to influence our bodies and behaviour, rather than our genes – and that these traits might be transmitted to our offspring.

What is the evidence for epigenetic inheritance? For plants, it is strong. For example, a particular flower shape found in some toadflax plants is faithfully passed on between generations, yet it does not appear to involve any difference in DNA sequence. This 'peloric' flower form, which has been known for more than 200 years, turns out to be caused by silencing of a gene through DNA methylation. However, there is no evidence to suggest this is 'adaptive'. In other words, the plant is not learning something in one generation that it commits to 'memory' epigenetically and then transmits to the next generation.

Evidence for transgenerational effects in animals is more scarce. The best example involves the effect of diet on an unusually coloured type of mouse called agouti. Normally, a litter of these mice has a range of coat colours from yellow to dark brown, thanks to the agouti gene. But if the pregnant mother is fed a diet high in certain vitamins and amino acids that are rich in methyl groups, she gives birth to more brown pups.

Another striking example is rat pups that are neglected by their mothers in the nest grow up to be skittish and timid adults. There is evidence that this is achieved through DNA methylation at a gene that regulates the response to stress. Methylation turns down the 'volume control' on the gene, resulting in permanent anxiety. According to the evolutionarily adaptive theory, this prepares the rats for a tough environment by making them more risk averse. This, however, is a within-generation effect, not one that is transmitted down the generations.

One (controversial) 2013 mouse study suggested that even the fear of a particular smell can be passed down epigenetically. Mice whose father or grandfather learned to associate the smell of cherry blossom with an electric shock became more jumpy in the presence of the same odour, and responded to lower concentrations of it than normal mice. And a 2016 study on frogs provides the strongest evidence yet that a father's lifestyle may affect the next generation. The research showed that sperm epigenetic tags change gene expression in embryos.

What about humans? Might the epigenetic consequences of starvation, neglect or disease be inherited across human generations? One study showed that people with a grandparent who went through a famine as a teenager died earlier, on average, if they were the same sex as the starved grandparent. The implication is that the experience of starvation changed the epigenome and that this effect was faithfully transmitted over two generations to compromise the health of the grandchildren. Studies of this kind are statistical and retrospective, so it is hard to investigate what is going on at a molecular level. Besides, it is difficult to rule out that these effects were transmitted through culture rather than epigenetics. Large-scale mapping of human epigenomes in relation to experience and disease is under way and may help resolve this conundrum. Even then it will be difficult to rule out the possibility that culture and behaviour within families is responsible for the transmission, rather than epigenetics.

### Evolution through epigenetics?

If individuals can acquire characteristics through interaction with their environment and then pass these on to their offspring, does this force us to rethink the gene-centred view of evolution and the idea of the gene as the basic unit of inheritance? A growing number of biologists think so (this is discussed in detail in Chapter 11). But not everyone agrees.

'The possibility that characteristics acquired during an individual's lifetime are epigenetically memorized and transmitted to the next generation has led to a revival in some quarters of the hitherto discredited theory of evolution known as Lamarckism,' says geneticist Adrian Bird of the University of Edinburgh. 'The consensus view, supported by a mountain of evidence, is still that evolution proceeds by natural selection among genetic variants that arise by accident. Much of the data is unreliable, but this has not stopped the spread of the idea that the environment "talks" to the epigenome and can ensure transmission of desirable (or, more often) undesirable traits without mutations. Nevertheless, epigenetic mechanisms could play some minor evolutionary role.'

In theory, what does Richard Dawkins, author of *The Selfish Gene* think? 'The "transgenerational" effects now being described are mildly interesting, but they cast no doubt whatsoever on the theory of the selfish gene,' he says. He suggests, though, that the word 'gene' should be replaced with 'replicator'. This selfish replicator, acting as the unit of selection, does not have to be a gene, but it does have to be replicated accurately, the occasional mutation aside. 'Whether epigenetic marks will eventually be deemed to qualify as "selfish replicators" will depend upon whether they are genuinely high-fidelity replicators with the capacity to go on for ever. This is important because otherwise there will be no interesting differences between those that are successful in natural selection and those that are not.' If all the effects fade out within the first few generations, they cannot be said to be positively selected, Dawkins points out.

## Is evolution predictable?

Evolutionary biologists have long debated whether rewinding the tape of life and replaying it would give similar results, or whether outcomes depend largely on chance events that push the course of evolution onto radically different tracks.

The two alternatives yield very different views of the history of life on Earth, with some prominent biologists, such as Simon Conway Morris, arguing that human-like, intelligent beings are inevitable products of evolution.

Others, such as palaeontologist Stephen Jay Gould, who popularized the tape of life metaphor, argue that if it were possible to turn back the clock, the history of life would not repeat itself. The world would be unfamiliar, and most likely lack humans.

Many studies have tested the reproducibility of evolution at the genetic level. In one, an international team took advantage of a natural experiment. Three different groups of terrestrial mammals have at some point in their evolution re-colonized the ocean, giving rise to what we now know as whales, walruses and manatees. Comparing the genetic changes in the three lineages, the researchers reasoned, should reveal whether evolution followed similar or very different paths in each case.

They sequenced the genomes of manatee and killer whales and bottlenose dolphins. The comparisons showed that many genes changed independently in each lineage, suggesting that randomness did indeed play an important role in their evolution.

But for 15 genes, natural selection led to exactly the same genetic changes occurring in all three lineages. This suggests that for some of the challenges of life in the sea, evolution repeatedly arrived at the same solution – that is, replaying the tape does indeed give much the same result again and again. This is a high-resolution replay of the tape, looking at what would happen to individual lineages, rather than what overall diversity would eventually result, which is what Gould looked at.

However, this result may say less about the predictable creativity of evolution than about a paucity of viable options. When the team performed a similar analysis of the genomes of dog, elephant and cow – related mammals that remained on land – they also found a comparable amount of convergence in their mutations, even though those animals share few similarities of lifestyle.

## Lack of options

This may imply that the vast majority mutations are lethal, so that evolution stumbles on the same few viable ones over and over again. Perhaps there is only so much that can be changed and still be functional.

Another study of the predictability of evolution looked at populations of the bacterium *Pseudomonas fluorescens* that rapidly diversify by mutation and selection into distinct types or 'morphs' when grown in test tubes. It found that when these mini-worlds are seeded with genetically identical microbes and the population size is large (around a billion cells per millilitre), each 'replay' results in highly similar patterns of evolutionary change. After just one week, *P. fluorescens* evolves into two new morphs called 'wrinkly' and 'fuzzy' spreaders. But this doesn't happen if the mutation supply rate is limited, reducing it by more than two orders of magnitude. Evolution only repeats itself if certain phenotypic innovations have a high probability of arising and are strongly favoured by selection.

If bacterial colonies that start out as identical clones evolve down different routes to reach a similar end point, what about colonies that start off as distinct? This has been studied using different *P. fluorescens* starter cultures with deleted genes encoding critical components of wrinkly spreaders and then allowing these 'defective' colonies to evolve. In all instances wrinkly spreaders eventually emerge by co-opting alternative genetic systems and structural components to bring about the necessary change. So, in the face of similar selective conditions, different lineages can find similar solutions to the same problems.

Replay life's tape, then, and while *Homo sapiens* may not evolve there is a high probability that introspective bipedal organisms with binocular vision will.

# 9
# The evolution of selfless behaviour

*Altruism is one of evolution's trickiest puzzles, and the question of why animals are so nice to one another is fiercely contested. If you accept that evolution is all about selfish genes, then what role has the group to play? Survival of the fittest means survival of the fittest DNA, which makes you and I mere vehicles in which our genes are hitching a lift on the road to posterity. Or maybe not ...*

## *The origin of altruism*

The theory of natural selection clearly explains how features such as the sharp teeth of the tiger, the thick fur of the polar bear and the camouflage of the moth evolved. Many social species, however, have traits that benefit others or the group as a whole, often to their own cost. It is much harder to see how such traits evolve. Darwin proposed that it was a result of the survival of the fittest groups, rather than the fittest individuals. **Group selection**, as this idea is known, has a long and complicated history and is one of the most controversial issues in evolutionary biology today. In 'What makes animals altruistic' (below), David Sloan Wilson, a leading proponent of group selection outlines the history of the field, and makes the case why group selection is a powerful force in evolution. There is also an interview with sociobiology pioneer and group-selection proponent Edward O. Wilson. Later in the chapter, we hear why other biologists disagree, and Richard Dawkins explains why his gene-centred view of evolution is sufficient to explain **altruism**.

## *What makes animals altruistic?*

*Many animals help other members of their group, often to their own cost. Evolutionary biologist David Sloan Wilson explains how such traits evolve, and makes the case for group selection.*

Wolves share food with other members of the pack. Vervet monkeys make alarm calls that risk attracting a predator's attention to themselves. Bees sacrifice themselves to defend a hive. But why? It is hard to see how such traits evolve by natural selection because individuals that behave in these ways would seem to have far less chance of surviving and producing offspring than more selfish members of the same group.

Darwin himself was acutely aware that the suicidal sting of the honeybee and most of the virtues associated with human morality, such as bravery, honesty and charity, posed a severe challenge to his theory. 'It must not be forgotten that ... a high standard of morality gives but a slight or no advantage to each individual man and his children over other men of the same tribe,' he wrote in *The Descent of Man* (1871).

The question, then, is how can traits that are oriented toward others or one's group as a whole – known as 'prosocial' traits – evolve when they appear to reduce the relative fitness of individuals within groups?

## A simple solution

Darwin realized that the problem could be solved if there was selection at a group level, not just at an individual level.

If groups composed of individuals who behave in a more prosocial way outcompete groups of individuals who behave in a less prosocial way, then traits that are for the good of the group will evolve. In short, natural selection between groups will counteract the costs of prosocial behaviour to individuals within groups.

Darwin's insight was the starting point for the modern theory of multilevel selection. According to this theory, biological systems are a nested hierarchy of units, from genes within individuals, individuals within groups, groups within a population and even clusters of groups.

At every level, the traits that maximize relative fitness within a unit are unlikely to maximize the fitness of the unit as a whole. Genes that outcompete other genes within the same organism are unlikely to benefit the whole organism. Individuals that

FIGURE 9.1 Groups that can work together successfully will have an advantage over other groups.

outcompete other individuals within their group are unlikely to benefit the group as a whole, and so on.

## A complex history

In the first part of the twentieth century, the idea of group selection was accepted by most evolutionary biologists. In fact, it was embraced rather too eagerly and uncritically. Influenced by the lingering idea that nature was the creation of a benign god, many biologists thought that nature was adaptive at all levels – what is good for individuals must be good for groups, and so on. As a 1949 biology textbook put it: 'The probability of survival of individual living things, or of populations, increases to the degree to which they harmoniously adjust themselves to each other and to their environment.'

While some biologists realized that the evolution of **prosocial traits** would often be opposed by selection within groups, it was frequently assumed that group selection would usually prevail – a position that has come to be known as 'naive group selectionism'. It was typified by British zoologist Vero C. Wynne-Edwards, who in 1962 proposed that organisms evolve to regulate their population size to avoid overexploiting their resources.

When the idea of group selection came under scrutiny in the 1960s, naive group selection was rightly rejected. But the backlash did not stop there. Based on arguments about how plausible it is, rather than actual experiments and studies, a consensus formed that between-group selection is almost invariably weak compared with within-group selection. As evolutionary biologist George C. Williams put it in his highly influential book *Adaptation and Natural Selection*: 'Group-related adaptations do not, in fact, exist.'

This was a major change in thinking, and the rejection of group selection came to be regarded as a watershed in the history of evolutionary thought. For the ensuing decades, most evolutionary biologists interpreted social adaptations as forms of self-interest that could be explained without invoking group selection. In his 1982 book *The Extended Phenotype*, Richard Dawkins went so far as to compare efforts to revive group selection to the futile search for a perpetual motion machine.

---

### Inclusive fitness and selfish genes

The rejection of group selection in the 1960s meant biologists had to come up with alternative theories to explain the evolution of social adaptations. Several were put forward, including **inclusive fitness** theory – also known as **kin selection** – selfish gene theory and evolutionary game theory.

---

The kin selection revolution started in a pub in the mid-1950s. Biologist J.B.S. Haldane was asked if he would give his life to save his brother. A few scribbled calculations later, he provocatively replied that he would only die for at least two brothers, or eight cousins.

Why? Because a gene coding for such altruism can only survive if it leaves enough copies of itself in relatives. Human siblings share on average half of their genes, and cousins one-eighth. Hence, two siblings, or eight cousins equal one self. This idea – that animals are more likely to show altruistic behaviour towards individuals they are related to – is called kin selection.

Haldane's colleague William Hamilton later drafted a mathematical description of the phenomenon, known as inclusive fitness, which assigns numerical values to the costs and benefits of an animal's actions. In theory, inclusive fitness makes it possible to calculate the extent of the spread of a given altruistic behaviour – such as staying with your parents to raise your siblings – through a population.

A revolution was ignited. Hamilton's maths has been used for decades by biologists studying cooperation in animals. Group selection mechanisms no longer appeared to explain altruism. The individual came to be seen as the protagonist of natural selection. Soon the gene took its place. Bodies could be regarded as merely the genes' way of making more genes, as famously expressed in 1976 by Richard Dawkins's metaphor of the selfish gene.

Yet the following decades saw a revival of group selection. What brought about such a remarkable change?

The rejection of group selection was based on the claim that, in practice, selection within groups always beats selection

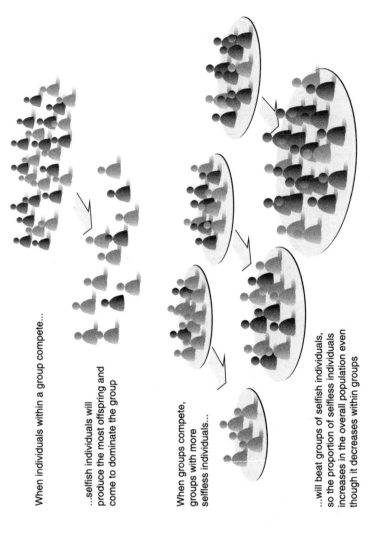

When individuals within a group compete...

...selfish individuals will produce the most offspring and come to dominate the group

When groups compete, groups with more selfless individuals...

...will beat groups of selfish individuals, so the proportion of selfless individuals increases in the overall population even though it decreases within groups

FIGURE 9.2 What happens when groups of selfish and selfless individuals compete?

between groups. However, studies have shown that traits can evolve on the strength of between-group selection, despite being disadvantageous for individuals within each group.

Take the water strider *Remigis aquarius*, an insect species that skates on the surface of quiet streams. Males vary greatly in their aggressiveness toward females, and lab studies show that within any group, aggressive males outcompete non-aggressive males for females.

However, the aggressive males also prevent females from feeding and can injure them, which results in a group with lots of aggressive males producing fewer offspring than groups with fewer aggressive males. Variation among groups is magnified by females fleeing groups containing lots of aggressive males and aggregating in groups with non-aggressive males. So these studies show that between-group selection is essential for maintaining non-aggressive males in the population.

In another experiment, a team of microbiologists grew *E. coli* in wells on plates. They then infected some of them with a virus, and mimicked the natural spread of viruses by using robotic pipettes to transfer them between wells. The team found that in some circumstances, a 'prudent', slow-growing strain of the virus was more successful than a 'rapacious', fast-growing strain. The rapacious strain often killed off all the bacteria in a well – and therefore itself – before it had a chance to spread. The prudent strain persisted for longer and so was more likely to get a chance to colonize other wells.

In this way, the prudent strain could remain in the population even though it was always outcompeted by the rapacious strain when both were present in a single well. In other words, it was only on the strength of between-group selection that the prudent strain survived.

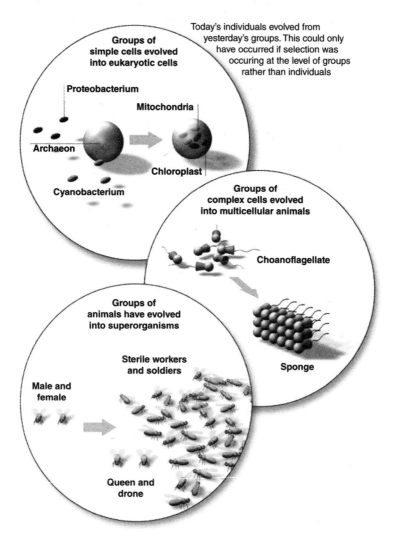

FIGURE 9.3 The evolution of today's individuals from yesterday's groups.

The conditions of this experiment closely resemble a scenario proposed by Vero C. Wynne-Edwards in the 1960s for the evolution of reproductive restraint in many species. Such restraints might not evolve in all species, but as this experiment shows, it is plausible that they can evolve in some species in some circumstances.

These two experiments involve very different spatial and temporal scales, but they embody the key problem and simple solution of group selection: the traits that benefit the whole group are not advantageous for individuals within the group and so require an additional layer of natural selection to evolve.

Darwin's problem is an unavoidable fact of life for all species, including our own: prosocial adaptations usually put individuals at a disadvantage relative to other members of their group. The only way for them to evolve is if there is another layer to the process of natural selection. That layer is group selection.

---

### Major transitions in evolution

Until the 1970s, evolution was thought to take place entirely on the basis of the accumulation of mutations over many generations. Then biologist Lynn Margulis proposed that complex cells did not evolve by small mutational steps from bacterial cells, but from symbiotic associations of different kinds of bacteria that became higher-level organisms in their own right.

In the 1990s, evolutionary biologists John Maynard Smith and Eörs Szathmáry proposed that similar major transitions occurred throughout the history of life, including the evolution of the first cells, the advent of multicellular organisms and the development of social insect

---

colonies (see Figure 9.3). They even suggested it could explain the origin of life, with groups of cooperating molecular reactions coming together to create the first life forms. The realization that evolution takes place not only by small mutational steps but also by groups of organisms turning into higher-level organisms represents one of the most profound developments in evolutionary thought. Today's individuals are yesterday's groups.

For a major evolutionary transition to occur, there has to be a shift in the balance between within-group and between-group selection. A group can only turn into an individual when between-group selection is the primary evolutionary force, and this in turn can happen only when mechanisms evolve that suppress selection within groups. The rules of meiosis, for example, ensure that all genes on the chromosomes have an equal chance of being represented in the gametes. If genes can't succeed at the expense of each other, then the only way to succeed is collectively as a group.

Major evolutionary transitions are rare events in the history of life but they have momentous consequences, as the new super-organisms become ecologically dominant. Eusociality in insects only originated about a dozen times – including in ants, bees, wasps and termites – but insect colonies comprise well over half the biomass of all insects.

These transitions are never complete as selection within groups is only suppressed, not eliminated. Some genes do manage to bias the rules of meiosis in their favour. Increasingly, cancer is studied as an evolutionary process that takes place within individuals, causing some genes to succeed at the expense of others, with tragic results for the group as a whole.

## Interview: From altruism to a new Enlightenment

*E. O. Wilson is the father of the field of sociobiology. In a 2012 interview in* New Scientist *he argued that group selection is the main driver of evolution. He is professor emeritus at Harvard University. Among his 25 books are the groundbreaking* Sociobiology *(1975),* Consilience *(1998), and the Pulitzer prizewinners* On Human Nature *(1978) and* The Ants *(1990).*

**In 2010 you were involved in a high-profile academic row over what drives the evolution of social traits such as altruism. Why should non-specialists care?**

Scientific advances are now good enough for us to address coherently questions of where we came from and what we are. But to do so, we need to answer two more fundamental questions. The first is why advanced social life exists in the first place and has occurred so rarely. The second is what are the driving forces that brought it into existence.

Eusociality, where some individuals reduce their own reproductive potential to raise others' offspring, is what underpins the most advanced form of social organization and the dominance of social insects and humans. One of the key ideas to explain this has been kin selection theory or inclusive fitness, which argues that individuals cooperate according to how they are related. I have had doubts about it for quite a while. Standard natural selection is simpler and superior. Humans originated by **multilevel selection** – individual selection interacting with group selection, or tribe competing against tribe. We need to understand a great deal more about that.

## How will a better understanding of multilevel selection help?

We should consider ourselves as a product of these two interacting and often competing levels of evolutionary selection. Individual versus group selection results in a mix of altruism and selfishness, of virtue and sin, among the members of a society. If we look at it that way, then we have what appears to be a pretty straightforward answer as to why conflicted emotions are at the very foundation of human existence. I think that also explains why we never seem to be able to work things out satisfactorily, particularly internationally.

## So it comes down to a conflict between individual and group-selected traits?

Yes. And you can see this especially in the difficulty of harmonizing different religions. We ought to recognize that religious strife is not the consequence of differences among people. It's about conflicts between creation stories. We have bizarre creation myths and each is characterized by assuring believers that theirs is the correct story, and that therefore they are superior in every sense to people who belong to other religions. This feeds into our tribalistic tendencies to form groups, occupy territories and react fiercely to any intrusion or threat to ourselves, our tribe and our special creation story. Such intense instincts could arise in evolution only by group selection – tribe competing against tribe. For me, the peculiar qualities of faith are a logical outcome of this level of biological organization.

**Can we do anything to counter our tribalistic instincts?**

I think we are ready to create a more human-centred belief system. I realize I sound like an advocate for science and technology, and maybe I am because we are now in a techno-scientific age. I see no way out of the problems that organized religion and tribalism create other than humans just becoming more honest and fully aware of themselves. Right now we're living in what Carl Sagan correctly termed a demon-haunted world. We have created a Star Wars civilization but we have Palaeolithic emotions, medieval institutions and godlike technology. That's dangerous.

**From all these big questions to the smallest creatures... I cannot interview the world's best known ant expert without asking: do you have a favourite?**

I do. It's an ant called *Thaumatomyrmex*. In all my travels, I've only seen three. They're very rare. It has teeth on jaws that look like a pitchfork. The teeth are extremely long, and when it closes the jaws, they overlap. In at least one species, the teeth actually meet behind the head. So what does this monster eat? What does it use those teeth for? I just had to know, so I sent an appeal out to younger experts in the field, particularly in South America, where these ants are found.

Eventually they discovered the answer: it feeds on polyxenid millipedes. These millipedes have soft bodies, but they're bristling all over like a porcupine. So the ant drives a spike right through the bristles and nails it. And what we hadn't noticed is that the ant also has thick little brushes [on some of its limbs], and members of the colony use these to scrub the bristles off – like cleaning a chicken – before dividing it up. That's my favourite.

## *Sparks fly over the origin of altruism*

To the average person, it might be hard to understand why biologists get so worked up about the origins of altruistic behaviours. But arguments about this topic are perhaps the most heated in science.

Since the evolution of altruistic traits such as forgoing reproduction in order to help raise another's offspring, has been explained by kin selection: the idea that helping your relatives – and therefore helping spread the genes you share with them – outweighs the cost of not having offspring of your own. It's the genes that matter, not the individuals in which they reside or the groups in which those individuals live. The idea was encapsulated in Richard Dawkins's pervasive metaphor of the selfish gene.

So when David Sloan Wilson and others revived the idea of group selection, it invoked a furious response from biologists in the gene-centred Dawkins camp. More fury was to come when, in 2008, E. O. Wilson wrote a paper outlining why kin selection was not the decisive factor in the evolution of sterile workers. He argued that once fully eusocial colonies (where a mother 'queen' is assisted at the nest by non-reproducing offspring) had developed, they continued to evolve by means of group selection, because groups that cooperate do better than those that do not.

This too, prompted a scathing response from many other biologists. Richard Dawkins, for example, in a comment article for *New Scientist* picked holes in Wilson's 'misleading' group-selection terminology, and argued that 'what really matters is gene selection' (also, see interview later in this chapter). He ended his article with these words: 'Evidently Wilson's weird infatuation with "group selection" goes way back: unfortunate in a biologist who is so justly influential.'

The arguments rolled on. In the 2010 meeting at the Royal Netherlands Academy of Arts and Sciences in Amsterdam, for example, one theoretical biologist denounced three of his colleagues as 'unscholarly' and 'transparently wrong', and wondered what could have led such 'talented, honest biologists' to be so 'misguided'.

The meeting was on the evolution of conflict and cooperation. The conference was just one stage of the controversy that had been raging over the work of three Harvard University scientists: mathematical biologists Martin Nowak and Corina Tarnita, and E. O. Wilson. The previous month they had published a paper in the journal *Nature* attacking inclusive fitness.

The details of their attack were technical and mathematical, but the consequences could be far-reaching. They said inclusive fitness is irrelevant to the real world and suggested replacing it with a series of equations that could describe the evolution of cooperation in far more detail than ever before. The problem, said Nowak and Tarnita, is that the calculations that describe inclusive fitness just don't work in the real world because they rely on a limiting set of conditions that nature does not stick to. Nowak pointed out that, in thousands of insect species, daughters leave the nest despite being as closely related to each other as the workers in an ant colony. This suggests there is some factor other than kin selection keeping workers in the nest and driving altruistic behaviour.

To some biologists this new model was a step in the right direction. To others it was lunacy, and more than 137 leading biologists signed a letter to *Nature* criticizing Nowak's paper in which they stated: 'Nowak *et al.* argue that inclusive fitness theory has been of little value in explaining the natural world, and that it has led to negligible progress in explaining the evolution of eusociality. However, we believe that their arguments are based upon a misunderstanding of evolutionary theory and a misrepresentation of the empirical literature.'

The arguments around group selection and inclusive fitness are still ongoing today.

## Interview: Selfish genes actually explain altruistic individuals

*Richard Dawkins has inspired millions with his popular science books, yet also drawn fire for controversial remarks – particularly on religion. In 2013 he spoke to* New Scientist *about group selection and how he felt about whether his role as the world's most famous atheist will eclipse his scientific legacy. He is emeritus professor of evolutionary biology at the University of Oxford.*

**Has what drives you changed over the years?**

It hasn't changed: a love of truth, a love of clarity, a love of the poetry of science. Insofar as I show hostility to alternatives, superstition and so on, it's because they are sapping education and depriving young people of the true glory of the scientific world view – I care especially about children in this case. It's tragic to see children being led into dark, pokey little corners of medieval superstition.

**One battle of yours has been against group selection – the idea that evolution works by selecting traits that benefit groups, not genes. You destroyed that paradigm, but then it came back again.**

Something else came back under the same name. If you look carefully it turns out to be things like kin selection rebranded as group selection. That irritates me because I think it is wantonly obscuring something that was actually rather clear.

I think part of why it came back is political. Sociologists love group selection, I think because they are more influenced by emotive evaluations of human impulses. I think people want altruism to be a kind of driving force; there's no such thing as a driving force. They want altruism to be fundamental whereas I want it to be explained. Selfish genes actually explain altruistic individuals and to me that's crystal clear.

**What subjects currently interest you in evolutionary biology?**

I'm fascinated by the way molecular genetics has become a branch of information technology. I wonder with hindsight whether it had to be that way, whether natural selection couldn't really work unless genetics was digital, high-fidelity, a kind of computer science. In other words, can we predict that, if there's life elsewhere in the universe, it will have the same kind of high-fidelity digital genetics as we do?

**When we are able to muck around with our own genes more, where do you think it will take us?**

The funny thing is that if you take the two parts of the Darwinian formula, mutation and selection, we've been messing around with the selection part with just about every species – except our own. We have been distorting wolves to Pekineses and wild cabbages to cauliflowers, and making huge revolutions in agricultural science. And yet with a few exceptions, there have been no attempts to breed human Pekineses or human greyhounds.

Now the mutation half of the Darwinian algorithm is becoming amenable to human manipulation, people have jumped to asking questions – what's going to happen when we start tinkering with genes? – while sort of forgetting that we could have been tinkering with selection for thousands of years and haven't done it. Maybe whatever has inhibited us from doing it with selection will do the same with mutation.

**Do you believe there is a genetic basis to irrationality?**

It would be very surprising if there wasn't a genetic basis to the psychological predispositions which make people vulnerable to religion.

One idea about irrationality that I and various other people have put forward is that the risks we faced in our natural state often came from evolved agents like leopards and snakes. So with a natural phenomenon like a storm, the prudent thing might have been to attribute it to an agent rather than to forces of physics. It's the proverbial rustle in the long grass: it's probably not a leopard, but if it is, you're for it. So a bias towards seeing agency rather than boring old natural forces may have been built into us.

That may take quite a lot of overcoming. Even though we no longer need to fear leopards, we inherit the instincts of those who did. Seeing agency where there isn't any is something that may have been programmed into our brains.

**If we are irrational, perhaps one of the reasons people bristle at you is they feel their nature is under attack.**

We accept that people are irrational for good Darwinian reasons. But I don't think we should be so pessimistic as to think that therefore we're forever condemned to be irrational.

**Would you rather be remembered for explaining science or taking on religion?**

To me they amount to the same thing – they are different sides of the same coin. But I suppose I'd rather be remembered for explaining science. I would be upset if people dismissed my science because of the religion.

## Taking it personally

The quest for a biological explanation for altruism is intricately linked to our ideas of goodness. No wonder that biologists have taken it personally.

The nature of altruism and its similarities to the human trait of goodness make it susceptible to political, philosophical and religious subjectivity. Studying the structure of an atom isn't personal: studying altruism can be. It certainly was for two figures in the history of altruism, Thomas Huxley and Peter Kropotkin.

Huxley, also known as 'Darwin's bulldog', outlined his thoughts on this topic in an 1888 essay entitled 'The struggle for existence': 'From the point of view of the moralist, the animal world is on about the same level as the gladiator's show ... Life [for prehistoric people] was a continuous free fight, and beyond the limited and temporary relations of the family, the Hobbesian war of each against all was the normal state of existence.' For Huxley, altruism was rare, but when it occurred, it should be between blood relatives.

Kropotkin, once a page to the tsar of Russia and later a naturalist who spent five years studying natural history in Siberia, thought otherwise. In Siberia he thought that he saw altruism divorced from kinship in every species he came across. 'Don't compete!' Kropotkin wrote in his influential book *Mutual Aid:*

*A factor of Evolution* (1902). 'That is the watchword which comes to us from the bush, the forest, the river, the ocean. Therefore combine – practice mutual aid!'

## Radically different conclusions

How could two respected scientists come to such radically different conclusions? In addition to being a naturalist, Kropotkin was also the world's most famous anarchist. He believed that if animals could partake in altruism in the absence of government, then civilized society needed no government either, and could live in peace, behaving altruistically. Kropotkin was following what he saw as 'the course traced by the modern philosophy of evolution ... society as an aggregation of organisms trying to find out the best ways of combining the wants of the individuals with those of co-operation'. He saw anarchism as the next phase of evolution.

Huxley was no less affected by events around him. Shortly before he published 'The struggle for existence', his daughter, Mady, died of complications related to a mental illness. In his despair over Mady's passing he wrote, 'You see a meadow rich in flower ... and your memory rests upon it as an image of peaceful beauty. It is a delusion ... not a bird twitters but is either slayer or slain ... murder and sudden death are the order of the day.' It was in the light of nature as the embodiment of struggle and destruction – the antithesis of altruism – that Huxley saw the death of his daughter and it was in that mindset that he penned his essay.

A suite of other fascinating characters would follow Huxley and Kropotkin. In the US there was the Quaker ecologist Warder Clyde Allee, who did the first real experiments on altruism in the 1930s and whose religious and scientific writings on the subject were often indistinguishable; in fact, he would often swipe text from one and add it to the other. Around the same time in the UK, J.B.S. Haldane, one of the founders of population genetics, was talking of altruism and kinship, and came close to developing a mathematical theory on the subject. But he stopped short – nobody quite knows why.

## The mathematics of altruism

A mathematical theory for the evolution of altruism and its relation to blood kinship would come a generation later with Bill Hamilton, who was both a passionate naturalist and a gifted mathematician. While working on his PhD in the early 1960s, he built a complex mathematical model to describe blood kinship and the evolution of altruism. Fortunately, the model boiled down to a simple equation, now known as Hamilton's rule. The equation has only three variables: the cost of altruism to the altruist ($c$), the benefit that a recipient of altruism receives ($b$) and their genetic relatedness ($r$). Hamilton's rule states that natural selection favours altruism when $r \times b > c$.

Hamilton's equation amounts to this: if a gene for altruism is to evolve, then the cost of altruism must be balanced by compensating benefits. In his model, the benefits can be accrued by blood relatives of the altruist because there's a chance (the probability $r$) that such relatives may also carry that gene for altruism. In other words, a gene for altruism can spread if it helps copies of itself residing in blood kin.

A generation of biologists were profoundly affected by Hamilton's rule. One them was the population geneticist George Price, an eclectic genius who became depressed when he came across Hamilton's work. He had hoped that goodness was exempt from scientific analysis, but Hamilton's theory seemed to demonstrate otherwise. Price went through the mathematics in the model and realized that Hamilton had underestimated the power of his own theory.

While working with Hamilton on kinship and altruism, the atheist Price underwent a religious epiphany. In an irony that turns the debate about religion and evolution on its head, Price believed that his findings on altruism were the result of divine inspiration. He became a devout Christian, donating most of his money to helping the poor. At various times he lived as a squatter; at other times he slept on the floor at the Galton Laboratory of University College London, where he was working. Price lived the life of the altruists that he had modelled mathematically.

## Does 'survival of the fittest' justify 'everyone for themselves'?

The phrase 'survival of the fittest' is widely misunderstood. Although the phrase conjures up an image of a violent struggle for survival, in reality the word 'fittest' seldom means the strongest or the most aggressive. On the contrary, it can mean anything from the best camouflaged or the most fecund to the cleverest or the most cooperative.

What we see in the wild is not every animal for itself. Cooperation is an incredibly successful survival strategy. Indeed it has been the basis of all the most dramatic steps in the history of life. Complex cells evolved from cooperating simple cells. Multicellular organisms are made up of cooperating complex cells. Superorganisms such as bee or ant colonies consist of cooperating individuals.

Besides, it is nonsense to appeal to the 'survival of the fittest' to justify any economic or political ideology, especially on the basis that it is 'natural'.

Is cannibalism fine because polar bears do it? Is killing your brother or sister fine because nestlings of many bird species do it? Just about every kind of behaviour that most of us regard as 'unnatural' turns out to be perfectly natural in some nook or cranny of the animal kingdom. No one can plausibly argue that this justifies humans behaving in the same way.

Yet even though such examples expose the utter absurdity of appealing to what is 'natural' to judge right from wrong – the naturalistic fallacy – we seem to have a strange blind spot when it comes to evolution. Survival of the fittest has been claimed to justify all kinds of things, from free markets to eugenics. Such notions still have a powerful grip in some circles.

However, natural selection is simply a description of what happens in the living world. It does not tell us how we should behave.

# 10

# *On the Origin of Species* revisited

To read On the Origin of Species *is to experience the extraordinary sensation of having a scientific genius enter your mind to guide you through his most important theory. To mark the 150th anniversary of the most influential piece of popular science writing ever published,* New Scientist *asked the geneticist, evolutionary thinker and author Steve Jones to summarize and update the book for the twenty-first century. Here is his updated version of Darwin's great work.*

# On the Origin of Species *for the twenty-first century*

Unique among scientific theories, evolutionary biology finds its roots in a popular book by a single author. The grey-bearded genius presented a new and radical view of existence: that life has changed over time and space, in part through a simple process called natural selection.

Charles Darwin called his work 'one long argument'. To a twenty-first-century reader it seems lengthy indeed, with only a single illustration to enliven its 150,000 words. But Darwin was a clear thinker and the book is an impressive piece of advocacy, moving from the familiar – how animals on farms have changed – to the less so, embryos and instinct included.

Darwin also shows how what might seem to be problems for his argument, such as the uncanny perfection of complex structures like the eye, are in fact part of the solution, and how apparent weaknesses in his case – the incomplete nature of the fossil record included – can easily be explained. Now and again he was wrong, as when, unaware of Gregor Mendel's work on genetics, he claimed that inheritance is based on the mixing of bloods, but mostly he was right.

Darwin described the process of evolution as 'descent with modification'. Today that might be rephrased as 'genetics plus time'. Offspring resemble their parents because they inherit DNA from them, but the copying process is not precise. Every round has errors, or mutations, and although they are individually rare – with perhaps one or two mutations in working genes each generation in humans – they can soon build up vast diversity. A copy of a copy is always imperfect, and for that reason alone, evolution is inevitable.

Darwin had a second insight. He saw that if a certain variant allows its carriers to survive, to mate and to pass on their heritage more successfully than others, in later generations it will spread. Such inherited differences in the chances of reproduction allow creatures to adapt to changing circumstances and can, in time, give rise to new forms of life. Natural selection, as he termed it, is a factory for making almost impossible things.

*On the Origin* was written in something of a rush. When Darwin discovered that Alfred Russel Wallace had hit upon the very idea he had been cultivating since soon after his return from the *Beagle* voyage, he condensed and refined his plan for a much longer book and set out to bring his theory to a wide audience. The book is far shorter than that first scheme, but as a result it is much clearer – which is, no doubt, why it made such an immediate impact. Darwin apologized again and again for leaving so much out and spent much of the rest of his life filling in the gaps.

If *On the Origin* was a hasty letter to its readers, the account that follows is no more than a postcard, but one that sketches out how Darwin might make his case today.

## Chapter One: Variation under domestication

*In which Darwin uses examples from domestication to explore the causes of variability and the principles of selection.*

Farmers have been unwitting evolutionists since they began, as they have shaped the characteristics of domesticated species. Nowhere is the power of human selection more clearly seen than by the fireside. Dogs were domesticated around 16,000 years ago in China, perhaps for meat. Their ancestors were wolves – and the two still share the same scientific name, *Canis lupus* – but dogs have changed mightily since then. The breeders were ruthless, killing off the animals of which they did not approve. This selective death, combined with selective sex, soon subdivided dogs into a huge variety of forms.

Some kinds have been distinct for centuries but most are less than 400 years old. In Darwin's childhood there were no more than 15 designated breeds. By the time *On the Origin* was published, the number had risen to 50. It is now around 400. Many of today's varieties have gained a distinct identity in no more than 30 or so canine generations. Sometimes a single mutation sparks a new variety.

The Irish wolfhound stands a metre high and weighs as much as 30 chihuahuas, but the difference in size between the two is due to a single gene, which occurs in one form in the large animal and another in the small. Most divergence between breeds, however, involves many genes, unwittingly selected by breeders aiming to refine the characteristics of particular lineages. Were the various breeds of dog to run wild, they might be classified by an enthusiastic but naive naturalist as separate species. How could a dobermann possibly belong in the same category as a chihuahua?

Even so, in spite of some logistical difficulties with sex, crosses among dog breeds produce fertile offspring, which is one of the several definitions of what a species is. In the same way, the European grey wolf crosses freely with domestic dogs, which is why it shares its name. The pedigree dog clubs have laid out the case for Darwinism in a few short decades.

'As for your doctrines I am prepared to go to the Stake if requisite... I trust you will not allow yourself to be in any way disgusted or annoyed by the considerable abuse & misrepresentation which unless I greatly mistake is in store for you... And as to the curs which will bark and yelp – you must recollect that some of your friends at any rate are endowed with an amount of combativeness which (though you have often & justly rebuked it) may stand you in good stead – I am sharpening up my claws and beak in readiness.'

*Darwin's supporter Thomas Henry Huxley, writing to Charles Darwin, 23 November 1859, regarding the* On the Origin of Species.

## Chapter Two: Variation under nature

*In which Darwin considers individual differences and highlights the wide degree of variability within species upon which natural selection works.*

The natural world is full of variation. The number of animal species identified so far is 1.8 million and there are no doubt many more yet to be discovered. Of other kingdoms of life even less is known. Many habitats are almost unexplored. A scheme to classify waterborne microbes sampled from the Atlantic, Pacific, Baltic, Mediterranean and Black Sea revealed thousands of new families of genes. They are proof of the presence of vast numbers of as-yet-unknown forms of life in a habitat that comprises 99 per cent of the biosphere.

Darwin was fascinated by the diversity of existence. He realized that new species are usually generated from the variation within existing forms which, genetics has revealed, is astonishing in its extent. Intra-species variability is scrutinized by natural selection, which promotes particular traits in particular environments. Should a creature occupy two distinct habitats separated by space or time it may become subdivided into two forms unable to exchange genes: that is, into two species.

Sometimes, the boundary between species is shifting and unstable, in testament to the dynamic and ever-changing pressures brought by a changing environment. In North America, for example, the red wolf and the coyote – distinct as they are – have lived in the same kinds of places for many generations. Each has maintained its identity. In recent years, though, they have been forced closer together by human activity, and have begun to produce fertile hybrids. For them, the barriers to gene-sharing were not quite complete, and what became for a time two species is sliding back to become one.

## Chapter Three: Struggle for existence

*In which Darwin describes the competition in nature for limited resources.*

All species have the potential to increase vastly in number, given time – but they do not, because of lack of food or because of disease, predation or for want of a home. In Darwin's day, the romantic poets and much of liberal society were in denial about this struggle for existence. Now we accept it, but we often forget quite how unforgiving it can be.

For example, around 400 million pet dogs live easy lives in human households. In the many cultures where dogs are despised, several million more roam in feral packs, forced to scavenge. Their days are chaotic, unhealthy and short. Meanwhile wolves, which once roamed throughout the world's northern forests, have been squeezed out by humans. Europe now has only a few thousand left and there would soon be none if they were not protected. Both wolves and feral dogs were unlucky to come up against the most intractable enemy that nature has yet produced; and both are paying the price in the struggle for existence.

## Chapters Four and Five: Natural selection and the laws of variation

*In which Darwin explains how inherited differences in the ability to survive and reproduce have shaped nature and the forces influencing the variation upon which natural selection works.*

Ten thousand years ago, much of North America was covered with ice. As the glaciers retreated they scooped out an intricate landscape of low hills, small lakes and rushing streams. Slowly, sticklebacks moved in from the sea. They found two distinct kinds of place to live in and adopted two ways of life. In lakes, where most of the food was available in open water, they became slim and active swimmers. In streams, where the shallow bottom was a better place to forage, they evolved to be robust and tough so as to grub around for food. Although they are not yet classified as distinct species, each form now prefers to stay in its own home and to mate with its own kind. Natural selection has worked to adapt the fish to the challenges they face, with only those individuals best suited to passing the test of survival getting the chance to hand on their genes.

As Darwin recognized, the entrance examination to the next generation has two papers. The first involves staying alive, but there is a second paper that turns on successful reproduction. It leads to what he called 'sexual selection'. Males of many species are forced to compete for the attention of females, and it pays a female to choose the best mate to father her offspring, the vessel of her own precious genes. This struggle for sex can lead to the evolution of bizarre structures such as the peacock's tail. Far from helping their bearers to stay alive, these sexually attractive traits are often a handicap. That may be why they work, for they indicate that those who bear them have what it takes to survive despite their costly sexual signal.

## Chapter Six: Difficulties on theory

*In which Darwin considers organs of extreme perfection and other apparent stumbling blocks for his theory.*

The mammalian ear is a wonderful example of how evolution can craft what at first glance looks like the work of an engineer. Fish can pick up sound waves with nothing more than simple pressure sensors on their bodies because water is an excellent conductor. When animals moved onto land, though, they needed to amplify sound waves, feeble as they are in comparison. Reptiles and birds have a lever – a tiny bone – between the outside world and the sensory cells of the ear that does the job, but in mammals the lever is comprised of three small, interlocking bones that work far more effectively.

The tale of the mammalian ear is one of make do and mend, and of structures modified for a new and different end. The earliest fish had no jaws. These developed later from bony arches that once supported the gills. Another gill arch became the single amplifying lever found in reptiles and their avian relatives. As the ancestors of mammals appeared, the ear began to hijack other structures. First, the position of the hinge between the upper and lower jaw shifted, freeing one bone of the upper jaw and one of the lower. These became the other two bones of the middle ear, so we hear, in part, with what our ancestors chewed with. Fossils reveal the whole process with a succession of creatures with more and more complete middle ears as proof of natural selection's ability to build on a series of successful mistakes and to craft complicated organs of apparent perfection.

Darwin knew nothing of genes, and his chapter on inheritance is the weakest in the book, but we now have molecular probes which reveal that the same genes are active in the fish sensory system as in the human ear – proof that genetics and evolution have become branches of the same science.

That trade-off between sex and death is hard at work in a certain Trinidad guppy. In some streams the males have bright orange spots that entrance the females. In others there lurks a nasty predator that spots the spots and eats those who carry them. Here, the males have evolved to be dull. But move them to a stream without enemies and in just a few generations the bright spots reappear, testimony to the power of sex.

## Chapter Seven: Instinct

*In which Darwin faces the issue of how behaviour might evolve.*

Dogs have long been used to hunt. Those with particular abilities – to track, to run, to recover corpses – were chosen as parents for the next generation. That history lives on in the instincts of today's breeds. Herding dogs such as border collies stalk sheep and do not bite them, but dogs bred to control larger animals, like corgis with cattle, go further and snap at their charges. Pit bull terriers are vicious creatures that can hold a bull – or a baby – by the nose. Can such psychological characteristics really evolve by selection?

Russian researchers in the 1950s showed that with strict selective breeding for tameness, once-wild silver foxes within a few generations began to wag their tails, bark and enjoy human company. Even their appearance altered, with new coat markings and floppy ears. After 30 generations they were completely tame. The brain, it seems, can evolve just as quickly as the rest of the body.

## Chapter Eight: Hybridism

*In which Darwin examines what keeps species apart.*

Species retain their identity because they live in different places, fail to have sex with each other or, if they do, produce infertile offspring. Hybrid sterility is due to genes, but because most species cannot be crossed it is hard to find out which ones are involved. Experiments with two kinds of Mexican fish show how little change is needed to keep species apart.

The platyfish lives in streams and is covered with elegant dark spots. In other streams nearby lives the swordtail, which looks much the same, although the males lack spots and have a long tail. The two species never cross in the wild, but can be persuaded to do so in the laboratory. Hybrids are spotty and survive reasonably well, at least in an aquarium, but if they are crossed in their turn to one of their own parents the next generation of offspring suffers a sinister fate. The small spots turn into lethal black tumours – and, in a macabre twist, males with these cancerous growths are much fancied by platyfish females.

A search through the DNA shows the problem. Cells have molecular brakes and accelerators that tell them when to divide, when to stop doing so, and when to die. In the hybrid fish, this control breaks down. The accelerator gene from one parental species refuses to respond to the command to slow down that comes from the other. As a consequence, the second generation hybrids die of cancer.

In an eloquent affirmation that evolution – shared descent – can unite quite distinct creatures, the overactive gene that leads to the problem in these fish is almost the same as the one which, when it goes wrong, causes skin cancer in ourselves.

## Chapters Nine and Ten: On the geological succession of organic beings

*In which Darwin considers the absence of intermediate varieties, explains why our palaeontological collections are full of gaps and describes how his theory can account for the pattern of succession from fossils to living forms.*

The geological record is like having just a few lines of a few pages of the history of the world. Much of it is incomplete, and because soft-bodied creatures are rarely preserved in the rocks some parts may be fragmentary at best. Since Darwin's day, however, vast numbers of fossils have been found across the world, and his concern that the geological record fails to support his theory now seems unduly pessimistic.

Darwin could only guess at the age of each stratum by esti-
mating the rate at which rock was worn away. Today we can
date fossils directly by examining how their chemical elements
break down with time. The first life has been traced back more
than 3 billion years and the death of the dinosaurs to 65 million years ago. Some of the records of that immense period are
impressively complete.

The Himalayas are full of fossils – not of mountain creatures
but of those of the sea – for long ago their peaks made up the
floor of the Tethys Ocean. The fossils include the antecedents of
great whales. The bones of the earliest ancestor of all cetaceans
are found in beds some 50 million years old. They belong to
a creature that had four legs and a tail, lived on the shore, and
looked a little like a seal. Its ears, though, had a unique structure,
now found only in whales.

The next prominent player, a million or so years later, 'the
swimming-walking whale', looked like a 3-metre-long otter.
Another million years on and the animal's nostrils had migrated
up the snout and the pelvis moved away from the backbone.
A further 5 million years saw the oceans inhabited by a long
mammal with tiny limbs.

Then came a great split. The ancestors of the blue whale and
its relatives – those that filter tiny creatures from the water –
began to develop gigantic sieves within their mouths, while
others retained the sharp teeth found in earlier whales and in
today's killers. More recent deposits reveal the splits between
dolphins and whales. A record that was once little more than
an enormous gap has, with infinite labour and some luck, provided a complete history of the evolution of the largest animals
that ever lived.

It has also given whales their rightful place – until not long
ago quite obscure – within the family of mammals. Their earliest

ancestors were close to those that gave rise to the hippopotamus. Whales, unique as they may seem, are hence members of a larger group that contains hippos, pigs, giraffes and cattle. DNA analysis backs up the record of the rocks. The whale's whole story has been revealed within less than half a century.

## Chapters Eleven and Twelve: Geographical distribution

*In which Darwin demonstrates the importance of geographical barriers and climate change to explain the distribution of life as we see it today.*

Penguins are charming birds with between 17 and 20 species, depending on which classification is used. They take a variety of forms, ranging from the statuesque emperor found in the Antarctic, to the fairy penguin of Australia, just one-twentieth its weight. Other penguins live in New Zealand, South Africa, South America and the Galapagos. They cannot fly (although the DNA suggests that albatrosses are among their closest relatives), so how did they reach such scattered places?

The oldest penguin fossils appear shortly after the demise of the dinosaurs around 65 million years ago. This ancestor of all living species lived in southern New Zealand and Marie Byrd Land, Antarctica, which were at that time separated by less than 1500 kilometres. Already the ancestral penguins had almost lost their wings. Fossils and DNA each show that the penguins' spread matches the advance and retreat of the ice. Starting around 35 million years ago, a series of ice ages made the Antarctic uninhabitable, and the birds retreated northwards as the glacier spread. In addition, icy currents swept some westwards, eastwards and closer to the equator.

After 10 million chilly years, the world warmed, cooled and warmed again, and the birds followed the retreating edge of the ice sheet into the far south. They left behind colonies scattered on cold-water shores across the globe. In the years since then, isolation and the challenges faced by each separate group have led to the diversity of penguins seen today.

## Chapter Thirteen: Mutual Affinities of organic beings: Morphology: Embryology: Rudimentary organs

*In which Darwin considers classification and shows how his theory can be used to organize the living world along evolutionary lines.*

As the past is the key to the present, so the infant is the key to the adult. Darwin realized that animals adopt the form which is adapted to their own evolved way of life as they develop. As a result, the deep relatedness between organisms might more readily be seen by comparing embryonic forms.

Darwin spent eight years working on barnacles, then an entirely obscure group but known to be related to crabs, insects and other jointed-limbed creatures. Adult barnacles vary in form from the familiar seashore kind to a sinister version that is an internal parasite of crabs and resembles a giant fungus. Whatever the divergence among the adults, though, the embryos are remarkably similar. They also resemble, to a lesser extent, the embryos of lobsters and crabs, and other jointed-legged creatures of the seas. All those in turn show affinities to the embryos of insects, a hint that butterflies have relatives on wave-battered shores, while that great group is in turn united in its earliest development with tapeworms and their relatives – creatures entirely different in their adult form. This shared identity, lost as the animals grow, shows how the embryo can reveal deep patterns of descent with modification, hidden as development goes on.

Darwin's own classification of life – of groups within groups as evidence of a shared hierarchy of descent that encompasses not just barnacles and butterflies but birds and bananas – turned only on what he could see with the naked eye, or down the microscope. Now genes – the units of evolution – have come to the rescue, and they reveal whole kingdoms of existence quite unknown only a few decades ago.

The new tree of life is a direct descendant of the single sketch of shared descent that appears in *On the Origin*. It is based on comparing the billions of DNA sequences now available from across the whole of existence. The affinity of whales with giraffes is a minor surprise when compared with the discovery of the relatively close affinity of all animals to mushrooms, and the emergence of a whole new kingdom of single-celled creatures, the Archaea, which have a structure and a way of life entirely different from the bacteria that they superficially resemble. They may even have cooperated with them, in the early days of life, to provide the nucleated cells that build all plants and animals of today.

## Chapter Fourteen: Recapitulation and conclusion

*In which Darwin expounds his 'long argument' and addresses the 'mystery of mysteries': why there are so many different species.*

A century and a half after *On the Origin*, Darwin can be seen to have been triumphantly right about almost everything. Evolution is now no more 'just a theory' than is chemistry and, like all other sciences, it provides a logical way of looking at the world. As he dared only to hope in that great book, light has been cast upon not just the worlds of plants and animals, but on ourselves and our origins.

Darwinism makes sense of what was once no more than a jumble of unconnected facts, and in so doing unifies biology.

Modern psychology, ecology and more find their birthplace in the pages of his greatest work. A century-and-a-half on, evolution is as central to our understanding of life as gravity is to the study of the universe. The closing words of *On the Origin* say it all: 'There is grandeur in this view of life, with its several powers, having been originally breathed into a few forms or into one; and that, whilst this planet has gone cycling on according to the fixed law of gravity, from so simple a beginning endless forms most beautiful and most wonderful have been, and are being, evolved.'

'The publication in 1859 of *On the Origin of Species* by Charles Darwin made a marked epoch in my own mental development, as it did in that of human thought generally. Its effect was to demolish a multitude of dogmatic barriers by a single stroke, and to arouse a spirit of rebellion against all ancient authorities whose positive and unauthenticated statements were contradicted by modern science.'

*Francis Galton, the English polymath and inventor, in his autobiography* Memories of My Life *(1908)*

# II

# The future of evolution

*What does the future hold for our understanding of evolution? In this chapter, biologist Kevin Laland argues that the concept of evolution needs a rethink. We also ask other leading biologists about evolution in the next 200 years.*

# Horfield Library
Tel: 0117 9038538
www.bristol.gov.uk/libraries

Borrowed Items 19/03/2018 15:24
XXXXXX6107

| Item Title | Due Date |
| --- | --- |
| * The food of love cookery school | 09/04/2018 |
| * How evolution explains everything about life : from Darwin's brilliant idea to today's epic theory | 09/04/2018 |
| * The bad mother | 09/04/2018 |

* Indicates items borrowed today
Remember to 'UNLOCK' your DVDs and CDs

Thank you for using self-service at Bristol Libraries.

## Evolution evolves: Beyond the selfish gene

*For more than 150 years it has been one of science's most successful theories, but does evolution need a rethink for the twenty-first century? Here, biologist Kevin Laland makes the case for an evolutionary upgrade.*

All scientific theories must incorporate new ideas and findings, and evolution is no exception. In recent years, our understanding of biology has taken huge strides. Advances in genetics, epigenetics and developmental biology challenge us to think anew about the relationship between genes, organisms and the environment, with implications for the origins of diversity and the direction and speed of evolution. In particular, new findings undermine the idea, encapsulated by the 'selfish gene' metaphor, that genes are in the driving seat. Instead, they suggest that organisms play active, constructive roles in their own development and that of their descendants, so that they impose direction on evolution.

Some biologists are trying to shoehorn the new knowledge into traditional evolutionary thinking. Others, myself included, believe a more radical approach may be required. We don't deny the roles of genetic inheritance and natural selection, but think we should look at evolution in a markedly different way. It is time for the theory of evolution to evolve.

Our current framework for thinking about evolution emerged only in the 1940s, with the integration of new knowledge about evolutionary processes and biological inheritance. This so-called modern synthesis is at the heart of how most people understand evolution (see Chapter 3). According to this view, the evolution of the features of an organism – collectively known as its phenotype – comes down to random genetic mutation, genetic inheritance and selection of those gene variants that bestow traits best adapted to the environment.

The modern synthesis has served us well: evolutionary biology is developing and thriving. But discoveries made over the past two decades are starting to reveal cracks in some of its central ideas.

## Not by genes alone

Take the notion that heredity happens via genes alone. In a classic nineteenth-century experiment, German biologist August Weismann cut off the tails of generations of mice, bred from the amputees, and found no reduction in tail length. This led to the view that genetic mutations in the germ line (eggs and sperm) are the only changes passed on to the next generation. But recent experiments suggest a more complex picture.

We now know that things other than genes are transmitted from parents to offspring. These include components of the egg, hormones, symbionts (microorganisms that live inside bodies), epigenetic marks (compounds that bind to DNA and turn genes on and off), antibodies, ecological resources and learned knowledge. At least some of these can lead to stable inheritance of phenotypes. For example, the transmission of epigenetic marks across generations is extremely widespread and, in plants, it can account for differences in fruit size, flowering time and many other traits. Epigenetic changes are often induced by changes in conditions within cells or the external environment, such as temperature, stress or diet, and unlike random mutations are often adaptive. Likewise, many animals inherit knowledge from their parents. Cultural inheritance occurs in hundreds of species, not just humans or vertebrates, but invertebrates such as bees and crickets too, creating similarities between even unrelated individuals.

These and many other findings suggest that the current focus on genetic mutations only captures part of the story of adaptive evolution – the slowly changing part. The broader view shows there are other ways to generate heritable variety. It also undermines the clean separation of development and heredity that Weismann's theory promoted. It is time to let go of the idea that the genes we inherit are a blueprint to build our bodies. Genetic information is only one factor influencing how an individual turns out.

And that's not all. We now also know that a given set of genes has the potential to produce a variety of phenotypes, depending on the environment in which the organism develops. This ability, called developmental plasticity, used to be dismissed as 'noise' or mere 'fine-tuning', but recent research suggests it may play a far more active role in the evolutionary process (see Chapter 8). As well as being able to respond in specific ways to particular conditions, organisms seem to have evolved the ability to respond flexibly to whatever conditions they experience. This adaptability results from a sort of Darwinian evolution occurring within organisms. It's as if each organism evolves as it develops, by generating new variation and selecting what works. This allows systems such as the immune system, nervous system and behavioural systems (through learning) to adjust to meet whatever environment the individual faces.

## Flexible phenotype

A flexible phenotype allows organisms to survive in the short term, and may then initiate evolutionary episodes – with genetic change following later. Consistent with this idea, several experiments reveal that organisms exposed to new environments develop characteristics that resemble those of closely related species adapted to these same environments.

FIGURE 11.1 The form a stickleback develops into can be influenced by environmental conditions.

For instance, marine sticklebacks reared on diets that are either benthic (bottom-feeding) or limnetic (mid-water) grow to resemble populations adapted to life in the corresponding environment. This suggests that adaptations may commonly arise through immediate responses to the environment, with natural selection favouring such individuals and subsequently cementing the useful features through genetic evolution.

There is also experimental evidence in insects, fish and amphibians that environmentally induced forms can evolve reproductive isolation, meaning that after a while they can no longer interbreed with other members of their species – a key step towards speciation. So developmental plasticity may play a critical role in both adaptation and speciation.

Features of development also undermine orthodox ideas about what factors influence the direction of evolution. The modern synthesis places natural selection in control, regarding it as the sole explanation for adaptation. Evolutionary biologists

have tended to think that evolution is not biased in any particular direction, since genetic mutation is assumed to occur
at random. However, this idea is challenged by 'developmental
bias' – the fact that certain characteristics can develop more
easily than others. This raises the intriguing possibility that the
diversity of life may not only reflect the survival of the fittest
but also the arrival of the frequent-est.

Developmental bias could help explain some fascinating
quirks of evolution. Consider parallel radiation, in which a
species in one location diversifies into several distinct forms
and, independently, the same diversification occurs in a different location. A famous example is cichlid fishes living in Lakes
Malawi and Tanganyika in Africa. Here many species exhibit
striking similarities in body shape with different species from
the other lake, despite being more closely related to species
from their own lake. These body shapes are adaptive, so natural selection has certainly been at play. But the forms we see
are not necessarily the only possible adaptive solutions. This
suggests there are features of cichlid development that make
certain forms particularly likely to arise. Developmental bias
could also help explain why cichlids – and some other groups
of organisms – are so diverse. It is perhaps because they are
particularly good at producing novel variants that can exploit
ecological opportunities.

This creative role for development contrasts with its traditional role of imposing constraints on adaptation. Constraints
explain the absence of evolution or adaptation, so have been of
limited interest. Many evolutionary biologists are now questioning whether this is the best way to think. Perhaps, rather
than merely setting limits on what forms are available for selection, developmental bias directs evolution by generating the
tramlines along which the engine of selection can proceed.

FIGURE 11.2 Why are some groups of organisms so much
more diverse than others?

## Not passive observers

There is a further way in which organisms might direct their
own evolution. Selection is portrayed as a process in which
external agents, such as environmental conditions, sort between
alternative variants according to their suitability. This is too passive. Organisms are not merely buffeted around by the forces of
nature; through their habitat choices and the way they modify
their environment, they play active roles in determining which
of their characteristics are useful. So they create some of the
conditions of their existence, and this influences their evolution.

By building a nest, which reduces temperature fluctuations,
for example, a bird weakens selection on the need for physi-
ological regulation of egg temperature, but creates selection for
refinements in nest design. Likewise, selection shapes a mammal
that digs a burrow less for ways to counter predation, and more

for resistance to fungal diseases. Such niche construction is not random, but systematic and directional. The animal manipulates the environment in a consistent, reliable way, to suit itself. In doing so, it biases the action of natural selection, imposing a direction on its own evolution, in much the same way that an animal breeder selects for particular traits in livestock.

Taken together, these discoveries challenge some of the fundamental assumptions of the modern synthesis (see 'Modern vs postmodern'). This new approach gives organisms a central role in their own evolution, and suggests that novel variation frequently begins not with mutation, but with changes in

FIGURE 11.3 By building a nest, a bird changes the selection pressures it faces.

phenotypes. It indicates that the direction of evolution does not depend on selection alone.

## Modern vs postmodern

Orthodox ideas about how evolution works are being challenged by new discoveries in genetics, epigenetics and developmental biology. This has led some researchers to propose that the current framework, known as the modern synthesis, be broadened into an extended evolutionary synthesis. The fundamentals remain the same, but they rest on quite different assumptions:

| Modern synthesis | Extended evolutionary synthesis |
| --- | --- |
| The major directing influence in evolution is natural selection. It alone explains why the properties of organisms are adapted to match those of their environments. | Natural selection is not solely in charge. The way that an organism develops can influence the direction and rate of its own evolution and its fit to its environment. |
| Genes are the only widespread system of inheritance. Acquired characters – non-genetic traits that develop during an organism's lifetime – are not inherited and play no role in evolution. | Inheritance extends beyond genes to include epigenetic, ecological, behavioural and cultural inheritance. Acquired characters can be passed to offspring and play diverse roles in evolution. |
| Genetic variation is random. Mutations that occur are not necessarily fitness-enhancing. It is mere chance if mutations give rise to features that improve the ability of organisms to survive and thrive. | Phenotypic variation is non-random. Individuals develop in response to local conditions, so any novel features they possess are often well suited to their environment. |

| | |
|---|---|
| Evolution typically occurs through multiple small steps, leading to gradual change. That's because it rests on incremental changes brought about by random mutations. | Evolution can be rapid. Developmental processes allow individuals to respond to environmental challenges, or to mutations, with coordinated changes in suites of traits. |
| The perspective is gene-centred: evolution requires changes in gene frequencies through natural selection, mutation, migration and random losses of gene variants. | The view is organism-centred, with broader conceptions of evolutionary processes. Individuals adjust to their environment as they develop, and modify selection pressures. |
| Micro-evolutionary processes explain macro-evolutionary patterns. The forces that shape individuals and populations also explain major evolutionary changes at the species level and above. | Additional phenomena explain macro-evolutionary changes by increasing evolvability – the ability to generate adaptive diversity. They include developmental plasticity and niche construction. |

There are two ways to view these new findings: we can try to incorporate them into the old framework, or we can extend the framework. Most evolutionary biologists take the first course, viewing plasticity and niche construction as being under genetic control, and seeing non-genetic inheritance as rare, unstable or functionally equivalent to genes. This view allows genes and selection to retain their explanatory prominence, at the price of downplaying new evidence. The alternative approach is to accept that the modern synthesis struggles to account for the new findings and to propose a broader alternative – an extended evolutionary synthesis. The two types of explanation can then be compared for their predictive power and ability to explain evidence, as well as their productivity in spawning new research questions and methods.

Evolutionary biologists who embrace this second approach are now acting on it. In 2016, an international consortium of 50 biologists and philosophers from eight universities announced a new research programme to investigate the evolutionary consequences of non-genetic inheritance, developmental plasticity and bias, and niche construction (see 'Time for change?'). These are exciting times for evolutionary biology, as the full ramifications of these ideas are explored rigorously for the first time. It remains to be seen whether our efforts will change the orthodox view. What is certain is that over the coming years, these advances will increasingly become the focus for evolutionary biologists.

My own view is that a new conceptualization of evolution is emerging. The selfish gene has proved to be a powerful and instructive metaphor, but the evidence now suggests it is misleading. Far from being master molecules, genes turn out to be just one of many channels through which cells respond to environmental inputs, and just one of several sources of heredity. Organisms are not the 'throwaway survival machines' envisaged by Richard Dawkins and others, but instead often take the lead in their own evolution, dragging genetic change along in their wake. Move over selfish gene, and make way for the orchestrating organism.

---

### Time for change?

A growing number of biologists believe we need to extend our ideas of how evolution works. This conviction rests on accumulating evidence that genes do not have sole control over development and heredity, and that organisms play active roles in their own fate and that of their descendants. These biologists have launched a wide-ranging research

---

programme to make the case for the so-called extended evolutionary synthesis (EES). One aim is to identify conceptual differences between the EES and orthodox thinking, and to test the distinctive predictions they make.

For instance, the traditional perspective sees biological novelty arising as a result of random genetic mutation, so it predicts that new forms are rarely advantageous. By contrast, the EES predicts that new forms are often adaptive because novelty commonly originates as a result of individuals adjusting to their environment as they develop. They will explore the extent to which this occurs, using a statistical analysis of published results describing how organisms respond to variation in environmental conditions.

Another group will focus on coral reefs to investigate the causes of biodiversity. Traditional thinking says that natural selection gives rise to organisms suited to diverse ecological conditions: the more different kinds of environments there are, the more species are expected to have evolved. The EES suggests that diversity also depends on properties of organisms – their evolvability. Organisms create their own habitats through niche construction, and they can also adjust to new conditions through developmental plasticity. Researchers will quantify how much of the diversity of coral reef fauna can be explained by the evolvability of corals, and how much by factors corals do not control.

The aim is to develop new ways to model the processes underpinning evolution. This will help us understand how the genes an organism inherits relate to the features it displays – that is, how genotype maps to phenotype. At the same time, philosophers and biologists will work together to update definitions of evolution, heredity and fitness.

## Evolution: The next 200 years

To mark the bicentenary of Charles Darwin's birth in 2009, *New Scientist* asked eminent biologists to outline the biggest gaps remaining in evolutionary theory. Here's what they said.

### RICHARD DAWKINS

Which facts about evolution had to be true, and which just happen to be true? Did the genetic code have to be digital in order for natural selection to work? Could any other class of molecules have substituted for proteins? How inevitable was the evolution of sex? Eyes? Intelligence? Language? Consciousness? Was the origin of life itself a probable event, and therefore is life common in the universe?

*Richard Dawkins is an evolutionary biologist at the University of Oxford*

### KENNETH MILLER

I don't think there are basic 'gaps' in the theory of evolution, which has proven to be a remarkably flexible scientific framework, brilliantly accommodating new data and even new fields of science, like molecular genetics. However, the most profound unsolved problem in biology is the origin of life itself. We know a great deal about the creative chemistry of the early Earth, but not yet enough to solve this problem.

*Kenneth Miller is professor of biology at Brown University, Providence, Rhode Island*

### FRANS DE WAAL

Why do humans blush? We're the only primate that does so in response to embarrassing situations (shame), or when caught in a lie (guilt), and one wonders why we needed such an obvious signal to communicate these self-conscious feelings. Blushing interferes with the unscrupulous manipulation of others. Were early humans subjected to selection pressures to keep them honest? What was its survival value?

*Frans de Waal is the Charles Howard Candler professor of primate behavior at Emory University, Atlanta, Georgia*

## ANDY KNOLL

Darwin explained how populations adapt to their environments, but Earth is a moving target, continually changing in response to both physical and biological forcings. The dynamic interactions between life and environment are imperfectly understood, but they underpin Earth's history and will determine the world our grandchildren inherit. The solution requires that we interject physiology – the interface between organism and environment – into analyses of how environmental change will affect life on Earth.

*Andy Knoll is the Fisher professor of natural history at Harvard University*

## STEVEN PINKER

How does selection leave its fingerprints on the genome? In particular, how does it work on the non-protein-coding parts, and what kinds of variation does it leave behind: a few common genes with small effects or many rare genes with large effects? This is necessary to understand how we differ from chimps and one another, and why we inherit diseases.

*Steven Pinker is the Johnstone Family professor of psychology at Harvard University*

## CHRIS WILLS

The biggest gap in evolutionary theory remains the origin of life itself. We now know that life began, probably near volcanically active zones, about 3.8 to 3.5 billion years ago, at a time when there was no free oxygen in the atmosphere. In the laboratory it has been possible to replicate such conditions and produce amino acids, primitive membrane-like structures and some of the building blocks of DNA and RNA. More recently, it has been found that, along with protein enzymes, RNA can catalyse chemical reactions, and it has even been possible to construct RNA molecules that can copy parts of themselves. But the gap between such a collection of molecules and even the most primitive cell remains enormous.

*Chris Wills is professor of biology at the University of California, San Diego*

## EÖRS SZATHMÁRY

Might evolution by natural selection explain complex thought? We know that a form of selection occurs within our brains as we develop and learn – synaptic connections and pathways that work well are reinforced, whereas weak ones deteriorate. But evolution also requires repeated replication and mutation to generate the variation upon which selection works to give adaptive solutions. At first glance nothing seems to replicate in brain tissue. Any search for neuronal replication will have to take place at a different level – perhaps in the patterns of connections between groups of neurons or in their patterns of activity. The idea is not so far-fetched. We already know that genetic evolution by selection is continuously shaping our immune response. If Darwinian dynamics can give us the flexibility to cope with new diseases, why not also the flexibility to find cognitive solutions to novel problems?

*Eörs Szathmáry is a theoretical evolutionary biologist at the Parmenides Foundation in Munich, Germany, and the Collegium Budapest, Hungary*

## STUART KAUFFMAN

Darwin changed our thinking as much as any scientist. Life, as zoologist Ernst Mayr said, only makes sense in terms of evolution. But major issues arise, such as the fact that Darwin did not know about self-organization. Abundant work over the past four decades has begun to show that self-organization plays a role, along with natural selection, in biology. One example is that lipids spontaneously form liposomes, the hollow bi-layered vesicles that must have yielded the cell membrane. Another is the spontaneous order in genetic regulatory networks, the understanding of which may lead to regenerative medicine and new cancer therapies.

*Stuart Kauffman is professor of biological sciences at the University of Calgary, Alberta, Canada*

## SIMON CONWAY MORRIS

'Evolution's biggest gap? Quite simple, old boy.' Professor Mortimer leaned back and grinned. 'Evolution equals change? Naturally, but that is only step one. What is life? A spectacular tightrope walk on a gossamer thread between vast regions of crystalline immobility and chaotic flux. If you don't like that metaphor, try thinking of a pack of cards a mile high with an elephant perfectly balanced on the top. And then there is its uncanny self-organization. Cells to consciousness – impressive, isn't it? Darwin got it right, and so did Newton. But then physics had Einstein. Perhaps now it is biology's turn.'

*Simon Conway Morris is a professor in the department of Earth sciences at the University of Cambridge*

# Conclusion

It is now more than 150 years since the publication of *On the Origin of Species*, one of the most important books ever written. In it Darwin outlined an idea that many still find shocking. He presented compelling evidence for evolution, and since his time the case has become utterly overwhelming.

Countless fossil discoveries have allowed us to trace the evolution of today's organisms from earlier forms. DNA sequencing has confirmed beyond any doubt that all living creatures share a common origin. Innumerable examples of evolution in action can be seen all around us, from the famous pollution-matching peppered moth to the emergence of diseases such as AIDS and H5N1 bird flu. Evolution is as firmly established a scientific fact as the roundness of the Earth.

So elegant is Darwin's explanation for how natural selection solves the problems of survival and creates the enormous diversity of life, that its outputs are still taken by many as those of an intelligent designer. Today, more than a third of American adults reject evolution entirely. Many others who accept evolution believe it is not the whole story and that it must be guided by a god.

For those who have never had the opportunity to learn much about biology or science in general, the claims about evolutionary theory made by those who believe in supernatural alternatives can appear convincing. This issue is more pressing than ever, now that evolution is gaining in importance not just as a matter of scientific rigour, but as a working technology. Most obviously, it is key to antibiotic resistance, which is shaping up to be a global crisis. And with the rise of personalized medicine, it is increasingly important to understand how genetics and inheritance interact with the environment.

That brings to the fore areas that are not yet part of the canon of evolutionary theory: epigenetics, for example, which studies how organisms are affected by changes in the ways in which genes are expressed, rather than in the genes themselves. Darwin's idea has already proven to be remarkably adaptable, accommodating the field of genetics, for example, but how it should be moulded to account for more recent discoveries is still a matter of debate.

And there are still gaps remaining in evolutionary theory. One of the biggest is the origin of life itself. Although we have made great progress in understanding the chemistry of the early Earth, we still don't know how the first life form arose from a soup of primordial chemicals. And what drove later explosions of evolutionary creativity?

In the next 150 years we can expect many of these gaps to be filled and an even more detailed answers to emerge for the eternal questions of how life began, and how we came to exist.

# Fifty evolutionary ideas

*The section helps you explore the subject in greater depth, with more than just the usual reading list.*

## Six places for evo-tourism

1 Any self-respecting evo-tourist should have the **Galapagos Islands** on their bucket list. These 19 islands – actually the tips of volcanoes – were where Darwin found inspiration among the mockingbirds, finches and tortoises.

2 For an equally exotic trip, try the **Moluccas** (also known as the Spice Islands). This series of islands, to the west of New Guinea, were where Alfred Russel Wallace devised his theory of evolution which came to him when he was ill with fever on the tiny spice island Ternate.

But if your budget doesn't stretch to tours of exotic islands, there are plenty more options:

3 **Down House** (Kent) is where Darwin and his wife lived for 40 years. It remains much the same as when they resided there. There you can walk the 'thinking path' that Darwin strolled along, and see the graffiti inscribed by his eldest son, William, in the schoolroom.

4 You can get a glimpse of Darwin's much earlier life by visiting his room at **Christ's College Cambridge**, where he studied between 1828 and 1831. The rooms were restored in 2009 to how they would have been in Darwin's time, and are now open to the public.

5 **Erasmus Darwin House** in Lichfield UK, is where Charles' grandfather Erasmus lived. Erasmus was also a renowned scientist (as well as physician and poet), and devised his own theories about evolution. His house, which Erasmus moved into in 1758, is now a museum.

6  **The Augustinian Abbey of St. Thomas** (Brno, Czech Republic) is where the monk Gregor Mendel carried out his famous plant breeding experiments, which led him to be posthumously dubbed 'the father of genetics'. Today the monastery houses a small museum dedicated to its famous former abbot, and visitors can also walk in the garden in which Mendel grew his plants.

## Five things named after....

1 **Wallacea** is a biographical region covering the Indonesian islands where many unusual species live, such as the dwarf buffalo and deer pig. This biodiversity hotspot is named after Alfred Russel Wallace, who spent eight years in this region.

2 Also in this region is the **Wallace line,** an invisible boundary in the ocean running between Borneo and Sulawesi, and between Bali and Lombok. It traces the path of a deep ocean trench and marks the boundary between Asian and Australian fauna.

3 The **Darwin Awards.** These tongue-in-cheek awards for 'outstanding contributions to natural selection through self-sacrifice', commemorate those people who, through their behaviour, remove themselves from the gene pool. But these are just one of the many things named after the great biologist, including one of the smaller Galapagos Islands and *Calceolaria darwinii* (also known as Darwin's slipper), an amazing orange-flowering plant he found in South America.

4 The ill-fated British Mars lander was named Beagle 2 after the HMS *Beagle* on which Darwin sailed to South America.

5 **F-test.** The name of this widely-used statistical test was coined in honour of the eminent biologist Ronald Fisher, who helped devise the 'modern synthesis' of evolution.

## *Eleven facts and anecdotes*

1 A Scottish fruit farmer called Patrick Matthew published the concept of natural selection well before Darwin and Wallace, in his 1831 book *On Naval Timber and Arboriculture*. Darwin later acknowledged this. An 1862 letter to Matthew opened: 'Dear Sir, I presume I have the pleasure of addressing the ... first enunciator of the theory of Natural Selection'.

2 Charles Darwin married his first cousin Emma Wedgewood. Although it was a successful union, Charles fretted that inbreeding may have been a factor in the early deaths of three of his ten children.

3 Before Emma accepted his marriage proposal, Charles had written a long list of pros ('My God, it is intolerable to think of spending one's whole life, like a neuter bee, working, working, & nothing after all') and cons ('cannot read in the Evenings – fatness & idleness – Anxiety & responsibility – less money for books').

4 Darwin used the original handwritten pages of *On the Origin of Species* as scrap paper for his children, who doodled on the back of them.

5 Darwin proposed a detailed theory of heredity called pangenesis that was completely wrong.

6 The phrase 'survival of the fittest' was coined not by Darwin but by the philosopher Herbert Spencer.

7 When Darwin first saw the Madagascan orchid *Angraecum sesquipedale* he noted its unusually long nectar reservoir, which would prevent most insects from gaining

access. 'Good Heavens what insect can suck it?' he wrote, and predicted the existence of an equally unusual pollinator, which had co-evolved alongside the orchid. This creature was discovered in 1903, 20 years after Darwin had died: the gigantic Congo moth, with a proboscis of up to 20cm long. But it was only in 1992 that the moth was actually seen feeding on the orchid, confirming the prediction Darwin made 130 years earlier.

8 The diaries of Emma, Darwin's wife, rarely mention her husband's work, but detail the constant stream of ailments that dogged her family. In particular they provide agonizing insights on Charles' poor health. Over the course of several months in 1840 Emma describes Darwin as 'exhausted', 'overtired + trembling', 'languid' and suffering 'great flatulence' (which then meant burping).

9 Alfred Russel Wallace's scientific reputation became somewhat tarnished by his speaking out in favour of spiritualism and mediums.

10 Wallace did, however, win a bet with a flat earth proponent. To win this wager, made in 1870, he devised an experiment along a 10km stretch of canal to demonstrate the curvature of the Earth.

11 Darwin's eldest daughter, Henrietta, was a key editor of some of her father's most famous works, but has been portrayed as a rigid believer who tried to supress the agnosticism in Charles' writings. However, her diary, written in 1871, deals with topics such as religion, free will and eternal life, and reveals that she had nuanced views on science and religion.

## *Three disasters, mishaps and coincidences*

1 Between 1848 and 1852 Alfred Russel Wallace collected an incredible variety of specimens on his expedition to Brazil. But on the voyage back to England the ship caught fire and Wallace lost everything except some sketches he had made. He swore never to travel again, but was off to Singapore just two years later.

2 After the monk and plant biologist Gregor Mendel died in 1884, the abbot who succeeded him at the monastery burned all Mendel's papers.

3 Darwin apparently only heard about the voyage of the *Beagle* by accident, via his uncle Josiah who learned of it from a doctor he was visiting about a 'buffy discharge from his bowels'.

## Five literary explorations of evolutionary ideas (and one not so literary one)

Many alternative evolutionary scenarios have been explored in the world of science fiction:

1 *The Committed Men* (1971) is M. John Harrison's first novel, and the influence of J.G. Ballard on its post-apocalyptic plot is obvious. What sticks in the mind most, though, is Harrison's playfully wrong evolutionary backstory: this is a book full of hopeful post-humans, evolving – never quite without reason – in the blink of an eye.

2 Philip K. Dick's short stories are full of wonders. The protagonist of 'The Golden Man' (1953) is sexually irresistible but essentially mindless – and of course his progeny are bound to succeed over poor brainy but pug-ugly humanity: intelligence is no defence against desire. Musical symphonies fed into 'The Preserving Machine' (1953), meanwhile, come out the other end as living creatures: all part of an experiment to see what happens when works of human manufacture are exposed to Darwinian selection. (A clue: it's nothing good.)

3 The current bible of the trans movement, Ursula Le Guin's *The Left Hand of Darkness* (1969), imagines a world where people are 'ambisexual', with no fixed gender identity. It's an engineered adaptation to a peculiar environment, and reveals Le Guin's grasp of that other, largely overlooked strain of evolutionary theory – the way cooperation can arise among communities and even between different species living in extreme conditions.

4 Evolution doesn't happen to individuals. Evolution happens to systems, and these systems don't have to be organic. For a coherent, persuasive and (70 years on) terrifyingly accurate glimpse into machine evolution, look no further than the Polish polymath Stanislaw Lem. The machine minds of his *Cyberiad* (1965) are far superior to our own but they've been carved by circumstance and contingency just as much as any piece of driftwood. Expect intellectual mayhem of the highest order.

5 Finally there is *The Time Machine* (1895) H.G. Wells. Historical commentators enjoy pointing out that this book is about the cruelties and injustices of industrialization. They're quite wrong. It is the most powerful, accurate-to-the-facts and pessimistic speculation on human evolution ever written. Evolution does not lead to excellence; nor, indeed, does it lead anywhere else. Pay especial attention to the epilogue: the last humans in H.G. Wells's London are exquisitely adapted to life on an Earth about to be swallowed up by its own sun. They've evolved into crabs.

And moving on from the science fiction genre to the realm of pure fiction ...

6 The 2012 comic animation *The Pirates! In an Adventure with Scientists!* (Aardman Animations) features a pirate captain seeking to win Pirate of the Year. When he captures Charles Darwin aboard the *Beagle*, Darwin discovers that the pirate's parrot is in fact the last living dodo.

## *Six experiments you can try at home*

Charles Darwin spent many years experimenting in his garden and house, to build up evidence for his theory of evolution by natural selection. Try your hand at eight of his classic experiments.

1   **Guided by the light**: Darwin noted that plant shoots emerging from the ground were sensitive to light and bent towards the sun as it crossed the sky. Keen to find out how the plants managed 'directed' movement, he and his son Francis experimented.

    You will need:

    - plant pots
    - soil
    - seeds (Darwin used canary grass, *Phalaris canariensis*)
    - aluminium foil
    - a lamp

Sow the seeds in pots. When the shoots emerge, switch on the lamp and watch them grow towards it. Repeat, but this time pop a foil 'hat' securely over the tip of the shoot before switching on the lamp.

Darwin worked out that the upper part of a shoot is necessary for the plant to respond to light: this paved the way for the discovery of plant hormones.

2   **Hungry plants:** Just after the publication of *On the Origin of Species*, Darwin was on holiday in the neighbouring county of Sussex when he stumbled upon the tiny insect-eating plant *Drosera rotundifolia,* or sundew. This inspired him to mount a series of detailed experiments to discover its preferred diet.

You will need:

- a sundew plant
- pretty much any food you like

Darwin tested the sundew's appetite for a vast array of different foods, including milk, oil, egg white, gelatin, sugar, hair, toenail clippings and even drops of urine. Make up your own morsels to discover what the plants like and dislike. Darwin's conclusion was that they were seeking out nitrogen.

If you can't get hold of a sundew, you might try experimenting with a venus flytrap to see what triggers it to snap shut.

3  **Dying young:** Darwin set up his 'weed garden' experiment in January 1857, and if you set about this in January or February you too should get good results.

You will need:

- a tape measure
- four pegs
- a hammer
- a ball of twine
- a spade
- a ball of garden wire snipped into 5-centimetre lengths

Use the pegs and twine to mark out a plot of lawn (1 metre by 0.7 metres) and carefully remove the layer of turf to expose the soil. Return every day to check for signs of germinating seeds, sinking a small length of garden wire next to each emerging shoot and collecting up (and keeping) wires where seedlings have died. In the summer, work out the extent of death on your patch. Darwin found that more than 80 per cent of his weedlings died young.

4 **A seed of an idea:** Back from the *Beagle*, Darwin pondered how plants and animals reached all the corners of the Earth. Conventional wisdom was that God put them there, but Darwin had other ideas. Perhaps seeds survived at sea and used ocean currents …

You will need:

- seawater (most pet shops sell salt water)
- glass jars
- your choice of seeds
- a sieve
- plant pots
- compost

Darwin used seeds of cress, radish, cabbages, lettuces, carrots, celery and onion. Label jars, fill with seawater and your seeds. After seven days, put the seeds in a sieve, rinse under a tap, and plant out in labelled pots. Darwin also studied longer periods in seawater, the effects of water temperature on germination, and whether seeds float. His experiments overturned the idea that seawater kills seeds. Of the 87 species he used, Darwin found almost three-quarters could tolerate at least 28 days in salt water.

5 **The price of competition:** In his 'lawn experiment', conducted in 1856, Darwin staked out a plot of old lawn and gave the gardeners strict instructions not to touch it. By mid-summer, once overgrown, it stood out in stark contrast to the lawn around it. '[O]ut of twenty species growing on a little plot of turf (three feet by four) nine species perished from the other species being allowed to grow up freely,' he wrote in *On the Origin*.

You will need:

- a tape measure
- four pegs
- a hammer
- a ball of twine

Hammer pegs into a patch of lawn to create a plot about 1 square metre. Wind twine around the pegs to make the plot abundantly clear to others. Get down on your knees and count the number of different plants growing in the plot. Count them again when thoroughly overgrown. In a competitive scenario, the tougher species will fare better than others.

6 **Ant attack:** Whilst at a spa in 1858, receiving treatment for one of his many ailments, Darwin performed a series of experiments on ant communication.

You will need:

- an empty jam jar or other suitable vessel for transporting ants
- ants from two different nests

Provided it's the right time of year (late spring and early summer), ants should be pretty easy to come by. Chances are you won't end up experimenting with the species Darwin stumbled upon (*Formica rufa*) but the important thing is that you locate two different colonies of the same species. Usher a few ants from colony A into a jam jar and carry them over to colony B. When Darwin performed this crude experiment, he found that the ants from one nest 'pitch unmercifully into a stranger brought from another'. From this, he correctly concluded that ants can sense and respond to chemical cues that differ between colonies. Repeat until you get a niggling feeling that you're torturing ants. You are.

## Three extracts from Darwin's letters

1 'My whole soul is absorbed with worms just at present!'

To William Turner Thiselton-Dyer, 23 November 1880, when Darwin was working on his final publication *The Formation of Vegetable Mould Through the Action of Worms.* Published the following year, just six months before he died, it shows that he retained an undiminished, almost childlike passion for his research.

2 'Poor Baby died yesterday evening. I hope to God he did not suffer so much as he appeared.'

To Joseph Hooker, 29 June 1858. Darwin was famously absent from the Linnean Society meeting of two days later when, in a hastily written paper, natural selection was formally presented to the world. His correspondence at the time is, however, dominated by a domestic crisis. Scarlet fever had hit his family, taking the life of his youngest child, baby Charles.

3 'If any man wants to gain a good opinion of his fellow men, he ought to do what I am doing: pester them with letters.'

To John Jenner Weir, 6 March 1868. In 60 years of letter-writing, Darwin badgered nearly 2000 people into exchanging more than 15,000 letters with him, many providing detailed observations of plants, animals and people from all over the world. His correspondents discussed both his ideas and theirs, helping to shape his published works. Today the letters are a window not only into Darwin's life and mind, but also into the lives of these, often otherwise unknown, collaborators.

## Ten places to find out even more

1 Go straight to the horse's mouth with *On the Origin of Species* (or to give it its full title: *On the Origin of Species by Means of Natural Selection or the Preservation of Favoured Races in the Struggle for Life*), Darwin's 1859 masterpiece. It's surprisingly readable.

2 The most famous of Wallace's 22 books is *The Malay Archipelago* (subtitle: *The Land of the Orang-Utan and the Bird of Paradise. A Narrative of Travel with Studies of Man and Nature*). Published in 1869 it describes his eight-year expedition to Malaysia, Singapore, Indonesia and New Guinea.

3 *The Selfish Gene* is Richard Dawkins' influential 1976 book that took evolutionary theory to a new level. He argued that an organism's urge to reproduce is prompted by its genes, which also direct it to favour its relatives, ensuring the survival of shared genes. The work, which has been translated into at least 20 languages and sold millions of copies worldwide, also introduced a now familiar cultural idea: the meme.

4 *Wonderful Life* by Stephen Jay Gould (1989). Evolution is not just about survival fitness, it's also down to Lady Luck. That was the controversial thesis put forward by Stephen Jay Gould in *Wonderful Life*. Using the Canadian fossil trove known as the Burgess Shale as the basis, he argued that chance played a role in which creatures became the forebears to modern life. The debate over his theory continues.

5 *The Double Helix* by James Watson (1968). The story of uncovering the DNA molecule's double-helix structure, told by one of its discoverers. Far from idealizing the scientific process, the book provides an honest account, sometimes painfully so, of the process of discovery.

6 For a more contemporary overview of genetics try the 2016 Pulitzer prize-winning *The Gene: An Intimate History*, by Siddhartha Mukherjee. After an account of the history of genetics, *The Gene* focuses on the past three decades of discovery in medical genetics, with examples from the author's family.

7 A single site for all of Darwin's published and unpublished writings can be found at http://darwin-online. org.uk/ including a major catalogue of his every publication and manuscript in the world.

8 The Darwin Correspondence Project (https://www. darwinproject.ac.uk/) finds and researches letters written by and to Charles Darwin and publishes complete transcripts together with contextual notes and articles.

9 See http://www.amnh.org/our-research/darwin-manu scripts-project to find a large collection of full colour, high-resolution images of faithfully transcribed Darwin manuscripts.

10 The writings of Alfred Russel Wallace, including the first compilation of his specimens, are at http://wallace-online.org/.

# Glossary

**Adaptation.** This is the process of change by which an organism or species becomes better suited to its environment.

**Amino acids.** These are the building blocks of proteins. The genetic code holds the blueprint for 20 different amino acids.

**Altruism.** In the context of biology, altruism means any behaviour which, at face value, helps the chances of survival of others at the expense of the altruistic individual. Honeybees, for example, that die when they sting intruders to protect their hives.

**Archaea.** These single-celled microorganisms make up one of the three major domains of life. Like bacteria, they lack a true cell nucleus and other complex cell machinery.

**Bacteria.** These single-celled microorganisms don't have a true cell nucleus or other complex cell machinery. They comprise one of the three major domains of life.

**Descent with modification.** Traits are passed down from generation to generation and sometimes undergo changes or modifications over time.

**DNA.** Deoxyribonucleic acid is the molecule that carries genetic instructions in most living things.

**DNA methylation.** The attachment of chemical methyl groups to genes, which alter the way they are expressed.

**Epigenetics.** This describes the array of molecular mechanisms that affect the activity of genes. Epigenetic 'switches' turn gene activity up or down, without altering the DNA sequence itself.

**Eukaryote.** This is an organism containing complex cells with a nucleus and other complex internal 'organs'. Eukaryotic organisms make up one of the three major domains of life, and include all animals, plants and fungi.

**Eusocial.** This describes social organisms such as honeybees, where a single female or caste produces the offspring and non-reproductive individuals cooperate in caring for the young.

**Gene.** A portion of DNA that acts as an instruction to make proteins.

**Genetic drift.** This is the process in which the frequency of a particular gene within a population changes due to chance alone, rather than selection.

**Group selection.** This is the idea that natural selection can act on whole groups of organisms, rather than just individuals.

**Inclusive fitness.** This explains how altruistic behaviour can spread through a population. It refers to the genes you share with close relatives, which are passed on to their offspring. It is why some animals have evolved to do things like help siblings raise their young.

**Kin selection.** This theory, based on the concept of inclusive fitness, holds that close relatives cooperate in order to perpetuate the genes they share.

**LUCA.** The Last Universal Common Ancestor of all living things.

**Modern synthesis.** This new understanding of evolution was developed in the 1930s and 1940s based on discoveries in many different fields. It framed the idea of evolution by natural selection in terms of genetics.

**Multilevel selection.** This theory holds that natural selection can operate at multiple levels at once – the individual, their family and the wider group.

**Natural selection.** This is the means by which tiny variations are naturally 'selected' by virtue of whether or not they help an organism to survive.

**Prokaryote.** This is a simple, single-celled microrganism which lacks internal 'organs' such as a nucleus. The prokary-otes are comprised of two groups: archaea and bacteria.

**Prosocial traits.** These are traits that are orientated towards others or one's group as whole, rather than the individual.

**RNA.** Ribonucleic acid is a molecule in all living things. Its main role is to act as a messenger to carry instructions from DNA for controlling the synthesis of proteins.

**Sexual selection.** This special type of natural selection oc-curs due to the preferences of one sex for particular charac-teristics of the other sex.

# Picture credits

All images © *New Scientist* except for the following:

**Chapter 1**

Figure 1.1: A portrait of scientist and naturalist Charles Darwin *c.* 1870. Julia Margaret Cameron/Underwood Archives/REX/Shutterstock

Figure 1.2: Galapagos Islands. Preserved birds at the Charles Darwin Research Centre, Santa Cruz. Gary Calton/Alamy

Figure 1.3: A portrait of Alfred Russel Wallace *c.* 1860. Hulton Archive/Getty Images

Tree silhouette c. Shutterstock/Waranon

**Chapter 2**

Figure 2.1: Peacock. Design Pics Inc/REX/Shutterstock

**Chapter 3**

Figure 3.1: Gregor Johann Mendel (1822–1884), Saustrian botanist and biologist. From 'Mendelism', London, 1905 by Reginald Crundell Punnet. Reginald Crundell Punnet/Universal History Archive\UIG/REX/Shutterstock

Figure 3.2: DNA (deoxyribonucleic acid). PASIEKA/Science Photo Library /Getty

**Chapter 4**

Figure 4.1: Boiling mud pot, Iceland. WestEnd61/REX/Shutterstock

Figure 4.3: Lightening. Steve Meddle/REX/Shutterstock

Figure 4.5: A first edition copy of Charles Darwin's *On the Origin of Species*. Nils Jorgensen/REX/Shutterstock

Figure 4.7. Source: *Nature Geoscience*, Vol. 1, p.49.

## Chapter 5
Figure 5.1: Insect eye, snipe fly. F1 Online/REX/Shutterstock
Figure 5.4: Portuguese man-of-war jellyfish at the Kimmeridge Marine centre. Steve Trewhella/REX/Shutterstock.

## Chapter 6
Figure 6.1: Buzzard (*Buteo buteo*), white morph, landing approach. Bernd Zoller/imageBROKER/REX/Shutterstock

## Chapter 7
Figure 7.1. Source: Michael Bell, Stony Brook University.
Figure 7.2: Straw-coloured Fruit Bat (*Eidolon helvum*), adult, in flight, Kasanka National Park, Zambia. Malcolm Schuyl/FLPA/imageBROKER/REX/Shutterstock
Figure 7.3: English Cocker Spaniel. Jan Tepass/imageBROKER/REX/Shutterstock.

## Chapter 8
Figure 8.1. Source: Eric Chaisson.

## Chapter 9
Figure 9.1: Winnowing wheat, Baaseli village, Rajasthan State, India. Environmental Images/Universal Images Group/REX/Shutterstock

## Chapter 11
Figure 11.1: Three-spined Stickleback (*Gasterosteus aculeatus*), Wateringen, Netherlands
Wil Meinderts/Minden Pictures/FLPA
Figure 11.2: Collection of beetles from around the world, Oxford University Museum of Natural History. Jochen Tack/imageBROKER/REX/Shutterstock
Figure 11.3: Cuckoo warbler. David Tipling/Solent News/REX/Shutterstock

# Index